Routledge Revivals

Religion and Biology

First published in 1922, this book represents an attempt to outline the biological approach to the questions of religious thought. The author posits the book as a contribution to religious thought in relation to the purpose of God in Nature, providing readers with an overview of the advances and changes in thought that had occurred in the years before the book was written. The examinations of the nature of man and of evolution in relation to religion make up the bulk of the book along with a look at the argument from beauty. The book will be of interest to students of religion, biology and philosophy.

Religion and Biology

Ernest E. Unwin

Routledge
Taylor & Francis Group

First published in 1922
by The Swarthmore Press

This edition first published in 2016 by Routledge
2 Park Square, Milton Park, Abingdon, Oxon, OX14 4RN
and by Routledge
711 Third Avenue, New York, NY 10017

Routledge is an imprint of the Taylor & Francis Group, an informa business

Publisher's Note
The publisher has gone to great lengths to ensure the quality of this reprint but points out that some imperfections in the original copies may be apparent.

Disclaimer
The publisher has made every effort to trace copyright holders and welcomes correspondence from those they have been unable to contact.

A Library of Congress record exists under LC control number: 23004744

ISBN 13: 978-1-138-65887-5 (hbk)
ISBN 13: 978-1-315-62050-3 (ebk)

RELIGION & BIOLOGY

BY

ERNEST E. UNWIN, M.Sc.

House Master and Science Master at Leighton Park, Reading

Author of "Pond Problems," "As a Man Thinketh,"

LONDON: THE SWARTHMORE PRESS LTD.

RUSKIN HOUSE, 40 MUSEUM STREET, W.C. 1

NEW YORK: GEORGE H. DORAN COMPANY

First published in 1922

Printed in Great Britain by
UNWIN BROTHERS, LIMITED, THE GRESHAM PRESS, LONDON AND WOKING

Nature! We are surrounded and embraced by her: powerless to separate ourselves from her, and powerless to penetrate beyond her. She is ever shaping new forms: what is, has never yet been; what has been, comes not again. Everything is new, and yet nought but the old. We live in her midst, and know her not. She is incessantly speaking to us, but betrays not her secret.

* * * * *

Everyone sees her in his own fashion. She hides under a thousand names and phrases, and is always the same. She has brought me here and will also lead me away. I trust her.

GOETHE'S APHORISMS,
translated by Huxley.

INTRODUCTION

This book is an attempt to outline the biological approach to the questions of religious thought. I have tried to write so that those who have not the time or opportunity to study the biological literature, may yet be able to appreciate something of the advance in knowledge and the change in thought during recent years; at the same time I believe I have a message for my fellow biologists and fellow schoolmasters.

At the end of each chapter a list of books is given for those wishing to go farther into the various problems. My facts can be verified by going to these books and to the biological text-books; but the implication of these facts is a personal matter and constitutes the purpose of this book. I have not tried to keep my own faith out of sight, for the best help we can give to one another is to make our individual contributions; and in dealing with God in Nature, my hope is that some may be helped to find Him not only revealed in Nature, but revealed also in Christ.

I am grateful to Nathaniel Micklem, M.A., the Editor of the Christian Revolution Series, for constant encouragement and helpful criticism. He has saved me from many obscurities; those that remain are my responsibility solely. The difficulties of the subject have crowded about me, as the book has grown under my hand; but I send it forth as a sincere contribution to religious thought in relation to the purpose of God in Nature.

Easter Vacation,
1922.

CONTENTS

		PAGE
	INTRODUCTION	7
CHAPTER		
I.	MAN A PRODUCT OF NATURE	11
II.	THE SPIRITUAL NATURE OF MAN	30
III.	THE EVOLUTIONARY CONCEPT	46
IV.	THE PURPOSE OF EVOLUTION	66
V.	DOUBTS AND DIFFICULTIES	88
VI.	THE APPEAL OF BEAUTY	122
VII.	THE KINGDOM OF MAN AND THE KINGDOM OF GOD	141
	A PRAYER	181
	INDEX	183

Religion and Biology

CHAPTER I

MAN A PRODUCT OF NATURE

"What is man that thou art mindful of him?"
HEBREW PSALMIST.

"And this to fill us with regard for man
With apprehension of his passing worth
Desire to work his proper nature out,
And ascertain his rank and final place;
For these things tend still upward, progress is
The Law of Life, man is not Man as yet."
BROWNING.

WE will begin our enquiry with man, and his place in Nature. This may seem a strange starting point, but there are reasons for the choice. In dealing with the facts of biology as a basis for argument and discussion, one is constantly coming up against the ignorance of the non-scientific person. We must attempt then to describe some of the basic ideas, and, because we are all tolerably familiar with the animal "man," it may be more interesting to use his structure to illustrate our thesis.

Then one is sometimes brought up short by the

refusal to accept for man any place in organic evolution. Some feel that man is a creature apart, specially created; though they might agree that the evolution-idea probably was true for the lesser creation, man, they feel, is outside the ordinary scheme of things. To take this view vitiates from the first all the arguments from the biological standpoint.

The refusal to consider man a product of nature may take subtle forms. A. Russel Wallace, a naturalist whose evolutionary studies demand close consideration, postulates a special creation for man's spiritual nature, while accepting the evolution of his bodily structure from lower animals. Others demand a special creation for man's self-consciousness. Others, again, think of the evolutionary method as a second attempt occasioned by revolt and fall. The danger of all such theories is, that, by removing man from the lap of Nature, we render insoluble the mysteries of his being and make unreal the problems of pain and evil.

"The denial of continuity is the only heresy evolution knows." I propose, then, to consider man and to place him in his right relation to the rest of Nature's work; indeed, to attempt to understand what the evolution idea is by considering the crown of Nature's work. The known is often a better starting place in a journey towards the unknown than to start with the simple forms.

* * * * *

In the early days of the nineteenth century a

series of treatises was published under the will and direction of the Earl of Bridgewater. One of the most interesting of these was the fourth of the series, published in 1832. It was written by Sir Charles Bell, a surgeon of repute, under the general title—" The Hand ; its mechanism and vital endowments, as evincing design, and illustrating the Power, Wisdom, and Goodness of God." The value of such a treatise is in no way destroyed by its ingenious and unconvincing arguments; indeed it is the aim of the present enquiry to establish the general underlying idea of these treatises. God in Nature? Yes; but not the mechanically working God of the Bridgewater days, but one who " has made things to make themselves, and to go on perfecting themselves—albeit they may be never separable in thought from Him," as Professor Thomson puts it in his recent Gifford Lectures.

As we turn the pages of Sir Charles Bell's volume we find it packed with information relative to the hands of man and vertebrate animals, forming valuable material for the argument from homologies put forward by Darwin.

This argument was summed up by Darwin in these words:

> The homological construction of the whole frame in the members of the same class is intelligible if we admit their descent from a common progenitor, together with their subsequent adaptation to diversified conditions. On any other view, the similarity of pattern between the hand of man, monkey, foot of horse, flipper of a seal, wing of bat is utterly inexplicable.

Let us examine the hand of man. We can see its covering of skin with nails and hairs. The skin is held tightly in some places, loosely in others, by an elastic tissue to the underlying parts. We can pull the skin away a little distance and when we let go it assumes its natural position. We can feel and see the movements of bones, muscles and tendons. We know from experience that there is blood in bloodvessels, and that nerves are present. It is quite simple to make out the number of bones and their arrangement. Each of the digits is hinged to a longer bone, the metacarpal, and these lying side by side form the body of the hand. One of these metacarpals, the one belonging to the thumb, has more freedom of movement, so that the thumb with its two phalanges can be opposed to the other fingers for grasping. The other four digits have each three phalanges. The bones of the arm can also be felt. The humerus running from the shoulder to the elbow, the radius and ulna from the elbow to the wrist. When the palm of the hand is facing upwards these two bones are almost parallel; the radius is on the thumb side, and the ulna, on the little finger side, extends to form the elbow knob. The individual bones in the wrist cannot easily be made out by feeling—there are two rows, each with four bones.

Having mastered the arrangement of bones in the arm and hand of man, it will be a simple matter to recognise that the common animals about us have the same general structure. The family cat or dog can be examined, and the pet rabbit too. It will require very little observation to realise that the arrangement of bones is the same. Some will be able to push the investigation further and add observations upon the hands and arms of birds, lizards, bats, pigs, horses and sheep. In some of these hands the modifications are very striking and suggestive in their relation to the different uses to which the hands are put; but we must not follow them in detail here (any textbook on comparative anatomy will supply the details for those who wish to continue the investigations).

Let us return to the hand of man. If the surface of the palm and fingers be examined, two kinds of markings can be seen. The creases formed by the flexing of the digits, and the smaller more regular ridges which show best upon the underside of the end segment of the fingers. These ridges have a definite pattern, consisting of concentric series of lines arranged about some central point. A set of these ridges can be made out on each finger-tip, at the base of the fingers there are signs of three sets, and one can be made out on each side of the palm. These last vary very much from individual to individual, for although all show some signs of the parallel ridges, only about one in twenty show the small centre around which the ridges run in concentric lines.

The hands of apes show very much the same arrangement, and the roughness caused by the ridges is useful in getting a grip when climbing. An examination of the cat's hand will suggest the right explanation. These concentric ridge-systems cover the vestiges of pads.

Our study has led us to an important point: the existence of vestigial structures in our hands. Now vestiges are clearly indicative of past history In such a word as "reign," the silent "g" is a vestige. Its value is merely historical, linking "reign" with "regnum," giving thereby a hint of the ancestry of "reign" and more than a hint of its meaning. In much the same way the vestiges in our hands tell of some use in the past, linking the present with an earlier stage in which the structures were of use. One can carry the observations a stage further. If the middle of the wrist be examined when the fingers are well extended, the great majority of people will notice a small cord below the skin. This is the vestige of the palmaris longus.

> In the arms of monkeys he would find (says Professor Keith [1]) this muscle well developed—and performing a very useful function. It must be remembered that the monkey uses its hand both as a hand and a foot. The palmaris longus acts on the palm, especially on the skin and pads of the palm, which are rough with papillæ to give firmness of grasp on the trees. The foot-like action disappears from the hand with the assumption of the upright posture, and hence in the anthropoid apes and in man the palmaris loses its function and becomes very small or may be quite absent.

I have taken the hand as an example and as an exercise in scientific observation. A similar story could be read in each and every part of the

[1] *The Human Body*, p. 86.

body of man. The skull, the leg and foot, the internal organs, the brain even, all support and substantiate this view that man has vestiges and homologies which are meaningless unless interpreted in the same way as those of other animals.

We have referred to the vestiges found in the hands of man, but there are many other vestiges of interest which give point to Walt Whitman's striking phrase, "Stuccoed with quadrupeds all over." The third eyelid, the external earshell with the muscles for moving it, the muscles for flicking the skin are vestiges with which we are all familiar, but there are others, muscles in relation to shoulder and pelvis, to arm and leg, which have been thrown out of use by the change from an arboreal life to life on the ground: at least the story told by these vestiges and the story revealed by the study of comparative anatomy are meaningless unless man has ascended from some kind of simple monkey.

Looking at the general structure of man it is clear that he is a Vertebrate, that with his warm blood, hairiness and the facts of his birth and suckling, he is a Mammal, and by reason of his skeletal features and in the details of brain and other systems of his body he is a member of the Primates

We might examine each system of the human body in turn, but it is unnecessary, for the anatomical evidence to be thus obtained would but bear out what we have already gathered. We see that man is a member of that group of mammals we have named the higher primates.

He is one of the three families included in that group. The central family is represented by the great anthropoids; man on the one side and the gibbon on the other represent the two other families. All three families we believe to have arisen from a common ancestral stock, but while the gibbon has clung to the ancestral form, man has progressively and aberrantly evolved to his present position. The great anthropoids have been steered in a middle course, and will without doubt at no distant date be extinguished when European civilisation reaches their jungle homes. Then will there be a wide gap between man and his nearest living allies.[1]

This extract is in the main correct, but the evidence is accumulating which puts the divergence of Homo much further back, indeed from the very simple Tarsius-like Primates, the basal stock from which New World and Old World monkeys and the apes have also sprung.

There are two ways in which biologists test their evolutionary theories. The one is by an enquiry into the fossil remains of animals. In the case of man the palæontological evidence is not so complete as it is in the case of many other animals, but there is already sufficient to give real support to the views expressed above, and there is every likelihood of this evidence becoming more complete as the remains of extinct forms are unearthed from the superficial layers of the earth's crust. Since the first draft of this chapter was written the Rhodesian remains have been discovered, and we must beware of dogmatism when new light is constantly flooding our enquiry. Professor J. Arthur Thomson's summary of our

[1] *The Human Body*, p. 51.

knowledge—based largely on Professor Arthur Keith's book, *The Antiquity of Man*—may be taken as a well-balanced statement.

During the Early Eocene ages, perhaps three million years ago, when grass was beginning to spread like a garment over the earth, there emerged an arboreal race of Mammals, the Primates, differentiated from other orders in digits and teeth, skull and brain. From this stock there diverged in the Eocene, first the New World Monkeys and then the Old World monkeys, leaving the main line (from our point of view) none the worse. Æons passed and the main stem, feeling its way towards the light, gave off in the Oligocene the branch of small apes (gibbon and siamang) and later the branch of large apes (gorilla, chimpanzee, and orang). This left, towards the end of the Oligocene (others would say in the Miocene), a generalised humanoid stem, perhaps weaker physically by the divergence of the apes, but probably dependent more on wits than strength.[1]

Then showing that the Java man, the Neanderthal type and the Piltdown man were probably offshoots from the main humanoid stem, Professor Thomson continues:

After the segregation of the branches represented by Pithecanthropus the erect, the slouching man of Neanderthal and Heidelberg, and the fine-brained Piltdown man, there was left the stem of modern man, which broke up in Pleistocene times into African, Australian, Mongolian, and European races. It is possible that the modern man type was distinguishable from collaterals a million years ago. If we mean by the antiquity of man the time since he reached what may be called the human standard in size of brain, Dr. Keith's conclusion is that this was reached by the commencement of the Pliocene period, which means

[1] *The System of Animate Nature*, p. 546.

over a million years ago. When the evidence of flints is considered, the tendency is to go further back still.

There may be errors in the conclusions of the authorities whom we have followed, and the estimates of time are very uncertain, but there is no great likelihood of errors which will affect the general impressions that alone concern us here. The antiquity of man is on a grand scale. There is a solemnity in the patience of the age-long manward adventure which has crowned the evolutionary process upon the earth. Three million of years ago the Primate stem sent out its first tentative branches, and the result was a tangle of monkeys; æons passed and the main stem, still probing its way, gave off the Anthropoids, which certainly rise to great heights. There was no pause, however, yet without hurry other experiments were made, and the terminations of these we know at Trinil and Heidelberg and Piltdown, for none of them lasted or was made perfect. Still the main line goes on evolving—and who will be bold enough to limit its insurgence?

The other way of testing our theories is by the study of embryology. One of the certain things that emerges from a study of man, or of any other animal that has been through many changes, is the wonderful way in which Nature economises in building material. When change of environment necessitates a change say in locomotion or in respiration, there will naturally be a disuse of some structure in the body of the animal. When this takes place it is very striking how economical Nature is, for old structures are utilised in the building of the fresh organs which are needed to meet the new conditions. We have all, at some time in our lives, watched the gradual metamorphosis taking place in the life cycle of the frog. We recognise the tadpole as the frog's larva, and are so familiar with it that we do not think how wonderful it is that an animal which breathes air and lives on land has a stage in its development which lives like a fish in the water. Let us revive Grant Allen's description of the scene.

> At the bottom of this shallow pond you may now see a miracle daily taking place, which but for its commonness we should regard as an almost incredible marvel. You may there behold evolution actually illustrating the transformation of life under your very eyes: you may watch a low type of gill-breathing gristly-boned fish developing into the highest form of lung-breathing terrestrial amphibian. Nay, more—you may almost discover the earliest known ancestor of the whole vertebrate kind, the first cousin of that once famous ascidian larva, passing through all the upward stages of existence which finally lead it to assume the shape of a relatively perfect four-legged animal. For the pond is swarming with fat black tadpoles, which are just at this moment losing their tails and developing their legs, on the way to becoming fully formed frogs.[1]

The tadpole breathes by means of gills like a fish, and has in its neck the skeletal supporting framework which is helpful in the efficient passage of water from the throat through the holes in its walls. It is by means of this water-current coming in contact with the fingerlike processes of the gills that the interchange of respiration is effected.

This branchial framework will not be required when the gills are replaced by lungs (the lungs are derelict hydrostatic bladders brought into a

[1] *The Evolutionist at Large*, p. 96.

new service). The gills disappear, the clefts close up, but in the change some parts of the framework are utilised to form a movable plate in the floor of the frog's mouth. The mouth is used as a pumping-chamber to fill the lungs, and this plate, operated by muscles, supplies the power. One pair of these passageways (gill-clefts) does not disappear. Each side of the frog's throat there is a shallow depression, into which a probe can be passed, leading out towards the surface of the head. The passage is blocked by the circular ear-drum, clearly visible on the side of the head behind the eye. Here is another story of change and economy. The ear, developed for use in water, is buried in the head close to the brain. This is no disadvantage, for vibrations in the water pass readily through the more solid material of the head, and are registered by the nervous apparatus of the inner ear, the real ear. Such an ear is at a great disadvantage when the animal climbs out on to the land, for the vibrations in the air need something more delicate than the side of the head to appreciate their presence. An ear-drum—a tiny skin-like plate—is used with a short bony rod to tap the vibrations against the bony skull, thus reaching the buried inner ear. But this drum, in order that it may not be thrown out of gear by changes in air pressure, must have air freely accessible to its inner side. This is made possible by utilising the first pair of gill-clefts.

I have described these details in the frog's development in order to make clearer the story in man.

It is during the third week that there appears on the neck of the embryo one of the most remarkable manifestations of a past stage of existence. On each side four grooves or depressions are formed. There can be no doubt that these represent the gill slits of fishes. We have every reason to suppose that the arches on the front and sides of the neck of the embryo represent the arches (i.e. the framework) which carry the gills, for into each one of them the aorta from the heart sends a branch, in the same manner as may be seen in the gill arches of fishes. No gills are actually developed because the placenta serves their purpose—that of respiration. Indeed, the clefts never open, but part of the first cleft remains, and forms the ear passage. Our ears are developed round the upper part of the first cleft. The lower jaw is developed in the first arch; the hyoid bone, on which the base of the tongue is set, is formed within the second and third arches. The cartilages of the larynx (Adam's apple)—the thyroid and cricoid, and also the cartilaginous rings of trachea (windpipe) and bronchi, are formed within the fourth and fifth arches. It is passing strange that we can recognise the same skeletal parts in our throats as we can see in the gill region of fishes. By the sixth week all outward appearance of gill-slits is lost; the manner in which the hinder ones disappear is instructive. In certain fishes the gills are covered by a lid or operculum, which grows back over them from the second arch, thus enclosing a gill chamber. In the human embryo a similar process is seen to occur. The third and fourth clefts are covered by a fold which grows backwards over them from the substances of the second arch, and as it grows back the clefts and gill chamber are obliterated. This is not always the case, however. The surgeon is frequently consulted by patients who complain of a discharge which escapes from a small opening situated on the one or the other side of the neck just above the inner end of the collar-bone. When a probe is passed within the opening it passes upward in the direction of the larynx for perhaps an inch or more. Such a fistula or opening represents the unclosed gill chamber. Occasionally other remains of gill-clefts are found as tags of skin or as auricular appendages on the upper part of the side or front of the neck.[1]

What Professor Keith describes in the case of man can be paralleled in the case of all air-breathing vertebrates. Reptiles, Birds and Mammals all exhibit at some stage in their embryonic development the relics of gill-clefts and their attendant structures. These soon disappear or are moulded into new parts. The heart, the bloodvessels and other parts of the body, also show this extraordinary embryonic metamorphosis.

Now these changes which man in his development undergoes, recapitulate in a very general way the earlier racial stages through which evolutionists, from the other evidence, describe as man's upward path. In the case of many lower animals which have become degenerate by reason of parasitism or of a sedentary life, the chief clue to their ancestry and classification is afforded by the embryonic or larval stages In man the embryological story confirms and fills out the evolutionary story. It is certainly a wonderful economy that is illustrated by the skeletal structures associated with speech and hearing. These two powers, so important in man's life, depend, for the nicety of their functioning, upon skeletal structures taken over from the gill-supporting framework of the long-passed fish stage.

There is one other way in which the story of the evolution of man, seen from this embryonic

[1] *The Human Body*, p. 98.

angle, can illustrate the same recapitulation idea. Many of the "diseases" or deformities found in man are the result of arrested development. Instead of a steady development of all parts of the body, an arrest may take place that results in the heart or the palate or the food-canal remaining in a more primitive state of development.[1]

These facts are inexplicable if man is a creature outside the general evolutionary scheme. What meaning can be given to the vestiges in his body, or to the structures which appear and disappear in his embryological development, or to these strange survivals which occasionally persist after birth, unless, to use Darwin's summing up at the end of *The Descent of Man*, "man still bears in his bodily frame the indelible stamp of his lowly origin"?

I have now indicated, in the barest outline, the kind of evidence brought forward by biologists to show what is meant by the ascent of man. For a fuller account, reference can be made to the books suggested at the end of this chapter. We can say then that man has arisen from simpler forms of animal life—this seems to me beyond doubt—but as we look at man we know that he is in many ways a creature apart, and the very fact that we are able to look at the problem from outside like this, convinces us that this is so. This uniqueness is not so much one of anatomical details, although these are evident, nor in the massive brain, although this has been perhaps

[1] See *Evolution and Disease*, J. Bland Sutton; *The Human Body*, Arthur Keith, chapters vii. and viii.

the chief factor in the establishment of his kingship in Nature, but in what we may call his powers.

> Some other creatures have words, but man alone has language—the power of expressing a judgment—which is *Logos*. Many other creatures have *intelligence*, which we can give a plausible account of in terms of perceptual inference, but man seems to stand alone in having *reason* or the power of conceptual inference. Many other creatures exhibit intelligent behaviour, which in a few cases may be controlled with reference to an objective end, as when the beavers dig a canal through an island in the river; but, so far as we know, it is only in man that behaviour rises into ethical conduct. Many animals are delightfully good, but only man is moral.[1]

Whence has man his characteristics? What factors have been operating in his ascent from some simple primate stock? We shall return to this problem later on in our enquiry, but suffice it here to say that we can only deal in probabilities; we cannot give a demonstrable proof of man's emergence, any more than we can give a demonstration of the evolution of birds or of insects. We cannot demonstrate evolution as we can demonstrate gravitation by some simple experiment like the fall of a ball released from our hands; but none the less we declare, without any doubt or qualification, that no other theory fits the facts, no other theory can hold the field in face of the evidences seen in the structure of plant and animal, and the records of missing links found in the rocks.

Every new discovery in biology, every fresh

[1] *The Bible of Nature*, J. A. Thomson, p. 186.

insight into the intricate web of life, serve to make more sure the fact of evolution, as the only interpretation of the way in which man and all animate nature have become as we know them to-day. It is well to insist on this, for so many people seem unable to distinguish between evolution as a modal interpretation of the world of Nature and the explanations put forward of its factors. When we ask how these things have come about—how birds have evolved from an ancient reptilian stock or insects from some worm-like form, or man and the apes from the simpler monkeys—we are confronted with very great difficulties. Evolution and Darwinism are not synonymous, and when Darwinism (an attempt to explain the factors causng evolutionary change) is enlarged and modified by fresh discoveries, and by greater insight into the workings of Nature, some people seem to think that the Evolution Idea is thereby discredited. Indeed, the reverse is the case. Every worker who has carried forward the great enquiry into causes and factors, which Charles Darwin so brilliantly inaugurated, has by his research and enquiry added fresh evidence in support of the way of Evolution.

Although we see man as the child of the past, we cannot say dogmatically thus and thus are the causes which have called him forth into his unique position. We know he is ancient; we know that at the time when he and the collateral lines leading to the monkeys and the anthropoids broke away from the ancestral stock, there were other creatures showing the same mutation of enlarged brains; we know that a larger brain spells educability. Life in societies favours prolonged infancy and intelligent co-operation, and these facts, together with the upright carriage which frees the hand for use as an instrument of progress, may have made it possible for man to rise head and shoulders above the other animals, into a fuller environment in which his life, under the influence of new stimuli, opens out into finer things. "A new door was opened when the foot became the supporting and branch-gripping member, and the hand was set free to reach upward, to hang on by, to seize the fruit, to hug the young one close to the breast. The evolution of a free hand made it possible to dispense with protrusive lips and gripping teeth, and thus there began the correlated enlargement of the brain-box and the bringing of the eyes to the front. Another arboreal acquisition was a greatly increased power of turning the head from side to side, and many other changes were involved in backbone and collar-bone, in chest and respiration, in hand and brain. 'It is the freed hand which is permitted to become the sensitive hand, which now, so to speak, goes in advance of the animal and feels its way as it climbs through life.'"[1]

But what can we say as to man's moral and spiritual nature? Have we here a clue to the whole matter? Is this the key to the mystery of creation? Man is certainly a product of Nature, he is no freak suddenly thrown into the

[1] *The System of Animate Nature*, p. 558.

world to sink or swim. We are here on the threshold of the whole problem, and it will be our endeavour to see what help can be brought from the biological side towards the understanding of a problem which leads into the realms of philosophy and religion. The end is an explanation of the beginning; and the beginning has in it the *entelechy*, the somewhat that bears the end in itself. Our purpose can be expressed in the following lines taken from the translation of an article by Prince Eugène Troubetzkoy: "Conceiving our life in this manner, the material evolution of the world becomes the incarnation and the expression of a spiritual meaning, of a divine event which is actually in process of coming to pass."

BOOKS

Darwin, C. . . .	The Origin of Species.
	The Descent of Man.
Drummond, H. . .	The Ascent of Man.
Keith, A. . . .	The Antiquity of Man
	The Human Body.
Huxley, T. H. .	Man's Place in Nature.
Wood Jones, F. .	Arboreal Man.
Lankester, E. Ray	The Kingdom of Man.
Thomson, J. A. .	Darwinism and Human Life.

CHAPTER II

THE SPIRITUAL NATURE OF MAN

Thou art the way.
Hadst Thou been nothing but the goal,
I cannot say
If Thou hadst ever met my soul.

I cannot see—
I, child of process—if there lies
An end for me,
Full of repose, full of replies.

I'll not reproach
The way that goes, my feet that stir.
Access, approach,
Art Thou, time, way and wayfarer.

ALICE MEYNELL.

IN speaking of man as unique, we mean, I take it, that he exhibits powers which we do not find in the lower animals. Man is aware of himself, he is self-conscious, and he is conscious of things which are not material. The fact that man in the contemplation of the beauty of Nature feels that he "cannot live by bread alone," and that he feels there is a somewhat within him which responds and reacts to the stimulus of beauty, lead to the postulation of a spiritual nature. It may seem a waste of time to some to attempt to establish this idea, for to them the reality of these spiritual faculties is as little doubted as the existence of any external reality we like to name—as the table at which they sit or the chair. I will quote a paragraph or two from *The Open Light*, by Professor Nathaniel Micklem,[1] in which this problem is treated in a fresh and striking way :

> Again, what man looking at St. Paul's Cathedral would say, " Why, it is only bricks and mortar and gilt ! " ? We are as sure as he who wrote " Si monumentum requiris circumspice " that " bricks and mortar and gilt " is not an adequate account of St. Paul's. If the Moonlight Sonata of Beethoven were only a series of noises drawn from a piano, it would be in no way superior to the braying of an ass or the screeching of a siren. If a man were in reality only the chymical particles out of which his body is made, then there would be no real difference between Christ and Barabbas. But if we judge neither architecture nor music nor persons in this way, we cannot so interpret the Universe. . . . But further, the mystery of matter lies not so much in the problem of what it is, but in the contemplation of what it will do. . . . Or consider again the " human face divine," the face of someone whose life and character we most admire. A physiologist may be able to tell us with much accuracy of what particles that face is composed, but that is a small part of the mystery ; how comes it that through the collocation and expression of those physical particles we can read the history of a lifetime, can see the whole lit up and transfigured by the spirit within, and can find in the depths of those eyes " the substance of things hoped for, the evidence of things not seen " ? The mystery of matter is that it expresses spirit ; it is unintelligible apart from its meaning. The printed signs that form this sentence, like the embossed figures of the Brail script, are an insoluble mystery apart from their meaning ; these alphabets were

[1] The Swarthmore Press (Christian Revolution Series).

created to express meaning. Nature likewise is a system of meanings, or it is an insoluble mystery. . . . Further, "matter" and "spirit" are both mysteries; but when we consider, we find that we know much more about spirit than about matter. We are conscious of being persons, or "spirits" to keep the old word; but of matter, though we cannot positively affirm that it is nothing apart from consciousness, we do and can know nothing except in relation to consciousness and mind.[1]

We will take it then that man has a spiritual nature. What is its origin? Was there a time in the past history of man when he had not got this spiritual nature? We must be careful in our use of terms. We are speaking of man's spiritual nature and naturally there was a time when there was no man to have this spiritual nature. In the same way we say in studying the evolution of birds that there was a time when these were no birds. But this is not to say that birds have been suddenly and specially created, for there is abundant evidence to show that birds are transformed reptiles and that feathers are glorified scales. The idea of continuity in the evolutionary story does not rule out large variations (or mutations as they are called); indeed, variations whether large or small form the raw material of evolution, they represent the note of experimentation. Birds are traced back to some reptilian ancestor, just as these ancestral reptiles can be traced back to a simpler vertebrate stock akin to the amphibians. Can we trace, in the same way, the evolution of man's spiritual faculties?

[1] *The Open Light*, p. 22 ff.

Let us think of the problem from the individual point of view. We are conscious, let us say, of our spiritual nature: How and when did it arise? At our conversion? Surely not, for conversion, whether it has been a sudden event, or a gradual growth, whether we are of the company of the once-born or the twice-born, or of the larger company of those who are a mixture of both, conversion is the personal realisation of our filial relation to and fellowship with God, and requires the voluntary facing of the right way. Without spiritual faculties we could not have realised our lack of harmony and fellowship; we may become more conscious of their presence and their value after conversion, but they are clearly not produced by conversion or as the result of conversion.

Do we get our spiritual nature at baptism? Surely not, for many who exhibit it have not been baptised, and baptism was originally a sign that a change had taken place and not the cause of the change.

Do they enter at our birth? What is birth? The act of being born is certainly the beginning of an independent existence, when seemingly a new environment opens up to us. Although there is superficially a pleasing simile in the infant's first intake of breath and the incoming of the breath of God, a moment's consideration will serve to dismiss this suggestion. The intaking of breath at birth, when the lungs are used for the first time, is not in essence different from what takes place when the three-weeks old chick draws its first breath from the air-space, gives a chirp of welcome, and chips its way out of the egg. In both cases it is a sudden change in the method of respiration that makes the birth or hatching a striking episode, but in itself it is merely an episode in a long story of development. The action of drawing air into the lungs for the first time is just as likely to cause the chick to become a spiritual person as the child. The more one knows of the facts of development in man or bird, the more clearly one sees the continuous process from the start to the adoption of an independent or free life and sees the embryo, although hidden within womb or egg-shell, as a living creature.

If then we do not become spiritual at birth, at what stage do we become so?

If we travel backwards in thought over our individual development, we can trace (in one unbroken line a story of moulding and elaboration from the far-off beginning in a single fertilised egg-cell. This wonderful development caused the old naturalists to put forward a theory of unfolding to explain the phenomena. The egg-cell was supposed to have the completed organism within it, and this was unwrapped during the embryonic period. We know now that the story is very different from that, and at the same time more simple and more wonderful. Every individual man starts—as do all other organisms above the simple firstlings which, like Amœba, are single-celled—from an egg-cell. This egg-cell is a single cell with its complement of protoplasm, containing the specialised part—the nucleus—which is the controller and leader in the segmental method of development, and the bearer of those qualities which cause the egg-cell of man to develop into a man and that of a crocodile into a crocodile.

The egg-cell divides into two, then into four blastomeres and by repeated divisions until a little cluster of cells is formed. The cells increase rapidly in number and group themselves into tissues, the tissues combine to form organs, the organs form systems, such as the blood system and the nervous system, and the systems together make up the complete organism. This developmental story, hidden away from sight in the womb, is paralleled in that of the chick within the egg-shell, or in that of the amphioxus in the warm, sandy bays of the Mediterranean, or in that of the sea urchin in the water around our rocky shores; in fact, a similar story, varying in length, can be related of all the Metazoans.

There does not seem to be any special place or time when the spiritual nature could enter in. What if it doesn't enter in? What if there is no need? What if it is in essence there all the time? What if we are thinking in terms of duality when we should really have a monistic conception? Our failure to find any definite place or time when a material body becomes a spiritual body may be due to the fact that there is no separation of the two, that "spirit" is inherent in "matter." I believe we shall find the answer to our problem at the very beginning and see matter and spirit united into the double-aspect or psycho-physical being.

Let us return for a moment to the development of man. We have already described in Chapter I some of the evidences which point to a general recapitulation of racial history in every individual development. We mentioned the "gill-cleft" stage with the fish-like heart. The study of post-natal development confirms this general idea, and, side by side with the growth of the body and the awakening of the mind, we can trace the beginnings of a moral nature and of the spiritual man. Is this individual story an epitome of the racial story? Does ontogeny give any indication of phylogeny? It does in the case of the tadpole. It does in regard to the bodily structure of man and of the other animals. Is it, therefore, too much to say that the moral and spiritual nature of man has evolved in the past in much the same way as they develop in the individual man?

Most of us have observed something of the post-natal development in a child; if we do not know of this as parents, we have ample material in the literature dealing with the development of mind, the study of instincts and behaviour in human life. Although we may not have kept a chart of the development of a baby's mind, and cannot say that here, when three weeks old, fear began to be shown and there the affections seem to start, we can all agree with the words of Henry Drummond:

> The most beautiful witness to the Evolution of Man is the Mind of a little child. The stealing in of that inexplicable light—yet not more light than sound or touch—called consciousness, the first flicker of memory,

the gradual governance of will, the silent ascendency of reason—these are studies in Evolution the oldest, the sweetest, and the most full of meaning for mankind. Evolution, after all, is a study for the nursery. It was ages before Darwin or Lamarck or Lucretius that Maternity, bending over the hollowed cradle in the forest for a first smile of recognition from her babe, expressed the earliest trust in the doctrine of development. Every mother since then is an unconscious Evolutionist, and every child a living witness to Ascent.[1]

From our knowledge as parents and schoolmasters, we know that this development of mind and of spiritual faculties depends very greatly upon the environment in which the child is placed. In some cases it is a question of the material environment, without in the presence or absence of fresh air, in the provision of healthy exercise and of suitable and sufficient food, within in the right functioning of the thyroid gland or the pituitary body. In other cases it is the mental and spiritual environment which is at fault, causing stunted or abnormal development. Although there is much more to find out by careful observation, both of normal cases and of abnormal cases, and especially where young children are removed from one environment to another, we know enough to say that there is in every normal healthy baby the potentials of a normal reasonable spiritual being. We can also say that these characteristics, which seem to mark man off so clearly from the lower creatures, are developed in the ordinary course of the unfolding human life. If this is so, and we exclude the theory of a second

[1] *The Ascent of Man*, chapter iv.

creation, there must be somewhat in the germ-cells of man—the material starting-place of the individual—which can eventually develop into these spiritual characteristics. These germ-cells were developed at a certain stage in the development of another individual, which in his turn developed from an egg-cell of the earlier generation, and so back and back to the simpler creatures from whose stock man has arisen.

The wonderful work in embryology, the investigations into the structure of the nucleus of the germ-cells, the finding of the seat of the hereditary characters in the chromosomes, add to our wonder as we consider the intricacies of development. We begin to realise that the germ-cell is a microcosm ; we begin to have some conception of the autonomous control and guidance towards a definite end, which is exhibited in this process of development as, amid the various stimuli, the form of a man is determined.

> What a piece of work is a man ! How noble in reason ! How infinite in faculty ! in form and moving how express and admirable ! in action how like an angel ! in apprehension how like a God !

Is there anything in the lower realms of animal life at all comparable to this autonomous and purposeful control ? If our ontogeny is to be any guide to phylogeny, we ought to be able to recognise some life-spirit, with some sign of this purposeful control of energy, even in the very simplest organisms.

It is here that the work of Hans Driesch can

help us. In his Gifford Lectures[1] Driesch states that the ultimate function of biology is the definition of life.

> Life is unknown to us except in association with bodies; we only know living bodies and call them organisms. It is the final object of all biology to tell us what it ultimately means to say that a body is "living," and in what sort of relation body and life stand one to the other.

Driesch is concerned to show the failure of the mechanicalists or materialists in describing life in physico-chemical terms. In this he succeeds. No one who has studied the examples which he brings forward, and has followed the careful reasoning which accompanies the facts, can doubt that there is somewhat in life which is not apparent in the domain of the inorganic.

He bases his discussion upon series of experiments, performed by himself and by other experimental embryologists, upon the behaviour of developing eggs and upon the regeneration of the parts of polypes and other simple organisms.

The egg of a sea-urchin divides into two blastomeres, these again repeatedly divide until a little ball of cells is formed. This is called the blastula stage. This blastula forms the larva of the urchin. Now if in the two-celled stage, one of the two blastomeres is killed or the two are separated, although each of them was going to form half of the larva, each develops into a complete and not into half a larva. Even one of the cells of the

[1] Hans Driesch, *The Science and Philosophy of the Organism*, 1908.

four-blastomere stage can develop into a complete larva.

Very ingenious experiments were performed in which the cleavage stages were altered, and yet the resulting organism was normal, showing that the organism could cope with very peculiar circumstances.

Suppose now that we allow the sea-urchin's egg to develop normally up to the blastula stage. In this stage it is a hollow sphere, the wall of which is a single layer of cells. It is similar all round, that is we cannot distinguish between top and bottom, right and left, anterior and posterior regions; but since it develops into a larva in which all these distinctions become apparent very soon, it must possess the three-dimensional mechanism, since the activity of the developmental process is going to produce different structures in each direction. Now the blastula, by very careful manipulation can be divided, cut into parts with a sharp knife. Since it is similar all round the direction of the cut is purely a matter of chance . . . really there are an infinite number of planes along which the blastula can be cut into two separate parts, and the direction of the plane is not a matter of choice, but purely a matter of chance. Nevertheless, each of the parts into which the larva is cut becomes an entire embryo. For a time the partial blastula—approximately a hollow hemisphere in form—goes on developing as if it were going to become a partial embryo, but soon the opening closes up and development becomes normal. It does not matter even if the two parts into which it is divided are not alike in size; provided that a part is not too small it will follow the ordinary course of development.

It is not possible to conceive of a machine being divided in any direction and yet remaining a whole machine.

Other experiments were performed upon polypes and planarians, and the organisms showed the power of restoration of the parts cut off from different places; the restoration was wonderfully adjusted to the circumstances of each case. This again overthrows the idea of a machine.

The extract given above was taken from a discussion of Driesch's work and thought by Dr. James Johnstone, in *The Philosophy of Biology.* He gives a faithful account of Driesch's argument in a short and a readable form. I propose to quote the concluding paragraph of the chapter entitled "The Vital Impetus."

We return to a consideration of the main results of experimental embryology in a later chapter, but let us notice here what is the direction in which these results, and those of the analysis of instinctive and intelligent action, carry us. It is towards the conclusion that physico-chemical processes in the organism are only the *means* whereby the latter develops, and grows, and functions, and acts. In the analysis of these processes we see nothing but the reactions studied in physical chemistry; but whenever we consider the organism as a whole we seem to see a co-ordination, or a control or a direction of these physico-chemical processes. Nägeli has said that in the development of the embryo every cell acts as if it *knew* what every other cell were doing. There is a kind of autonomy in the developing embryo, or regenerating organism, such that the normal, typical form and structure comes into existence even when unforeseen interference with the usual course of development has been attempted: in this case the physico-chemical reactions which proceed in the normal train of events proceed in some other way, and the new direction is imposed on the developing embryo by the organisation which we have to regard as inherent in it. This same direction and autonomy must be recognised in the behaviour of the adult organism as a whole. What is it? We attempt to think of it as an impetus which is conferred upon the physico-chemical reactions which are the manifestations of the life of the organism. It is the *élan vital* of Bergson, or the *entelechy* of Driesch.

Driesch postulates an *entelechy* to explain these facts of morphogenesis. It is a " somewhat " in living matter which tends to produce completeness and perfection out of incoherence and imperfection. An analogy is used of a workman who may, with his pile of bricks and mortar and the same expenditure of energy, build a wall, an archway or a small house. These different results may arise from the same initial arrangement of materials and energies, because there is something more than energy and matter present. There is the intelligence of the workman. We can compare *entelechy* to this controlling intelligence. I am not concerned here to support to the full the very interesting contribution of Driesch to the problem of vitalism. That he makes out his case for a " somewhat " in life which, as a vital principle, acts as a controlling force, and which in its simplest expression is simply morphogenetic (i.e. dealing with the development of the body), seems to me incontrovertible; but he gives one the impression that this *entelechy* is something operating upon organisms, something separate and distinct from matter. It seems more likely that the organism is a psycho-physical individual. It is, " while it lives, an indissoluble psycho-physical being. The mind and the body are both abstractions, very convenient for purposes of discourse; there is but one reality, the life of the organism, which has a subjective aspect known as psychosis and an objective aspect known as neurosis. The living creature gives an account of itself in two ways. It can know itself as something extended and intricately built up, burning away, moving, throbbing; it can also know itself as the seat of sensations, perceptions, feelings, wishes, thoughts. But there is not one process, thinking, and another process, cerebral metabolism; there is a psycho-physical life—a reality which we know under two aspects."[1]

This view, summarised by Professor Thomson, seems to get over the difficulty of discontinuity in Nature, and yet is quite clearly on the side of the vitalists in their thesis of the uniqueness of life.

To what conclusion, then, are we led? I believe, as a result of our enquiry, that we are led to the view that there is no break in the scheme, and no sudden incursion from without. It may seem a stupendous assumption to make upon so little evidence, for I have not attempted to do more than give the briefest outline of the line of thought, yet, having reviewed the various alternatives, I am convinced that the spiritual nature of man is the real nature of man, and that there has not been a sudden incoming of the Holy Spirit, or even a gradual infiltration; rather, I believe, that his spiritual nature is the development from and the outcome of the essential spirituality of the living organism. We can hear, as we listen

[1] *The System of Animate Nature*, p. 247.

to the evolution story, "an onward-advancing melody"—to use Lotze's expressive phrase. In the firstlings of life, along with those rudimentary characteristics from which we trace the gradual elaboration of contractility, assimilation, respiration, and all the other phenomena of the living cell-life, there is this dim and diffuse spirituality, or consciousness if that term is preferred. At first this is only seen in its morphogenetic rôle, but, gradually, this dim consciousness through the age-long adventure of life becomes concentrated or developed into that kind of consciousness which we recognise as intelligence—a more obvious, purposeful control of the factors of life. The next step in the story is the rise into self-consciousness, which we recognise as one of the characteristics of man, and this self-consciousness develops into the higher stages of this long and slow ascent, and these last stages are called spiritual.

We have in this chapter jumped over many difficulties and obscurities, some of which we shall consider in later chapters; but as an outline it is, I believe, truthful to the facts, and its main idea is supported by the research and thought directed to the problem by the biologists from below and by the philosophers from above. Man is a spiritual being. His spiritual nature, like his bodily frame, has been developed from the far-off prophecy in the firstlings of the Universe.

In insisting upon the continuity of Nature, men of science have been better theologians than the theologians themselves. If God exists at all He is the God of all

Nature and of every natural law. There are no gaps in His workmanship, no breaches of continuity in His activity. All Nature is an activity of His, and every natural law a principle of that activity.[1]

BOOKS

DRIESCH, HANS . . The Science and Philosophy of the Organism. Gifford Lectures.

JOHNSTONE, J. . . The Philosophy of Biology.

MICKLEM, N. . . The Open Light.

THOMSON, J. A. . The System of Animate Nature, Gifford Lectures, chapters iv. and v.

[1] *The Study of Nature and the Vision of God,* G. J. Blewett.

CHAPTER III

THE EVOLUTIONARY CONCEPT

"The nature of a power at work in any process is only revealed in the process as a whole."

PROFESSOR A. S. PRINGLE-PATTISON.

IN the two preceding chapters we have staked out the ground and given some indication of our purpose. We have tried to get a glimpse of the evidence upon which biologists base their belief in the ascent of man, and have indicated the kind of arguments for the autonomy of life put forward by those biologists who are concerned with the philosophy of biology.

In this chapter I wish to discuss the evolutionary concept, in order to remove some of the difficulties which have gathered around the idea. We shall then be more ready to travel forward towards some attempt to understand the purpose of God.

Man is a part of Nature. His arrival on the earth has been a "becoming"; he is a product of the scheme just as certainly as the arrival of insects or birds is a product of the working of the scheme. The evolutionary method has to be judged by its success in producing man, not by the production of insects or birds; neither must we judge it by man's failure in understanding it, nor by his unsuccessful attempts in controlling it.

We stand, as spectators, watching the great drama of evolution, and we endeavour to grasp the plot. We have arrived too late to see the early acts and scenes, we get some indication, by a hint here and there, of what has gone before, and each actor, as he comes on the stage, gives a very rapid and very general summary of the preceding acts as they apply to him and to his place in the drama; but many of the subsidiary plots have been worked out before our arrival. The difficulties of the spectators are thus very great. Then we must remember that the plot cannot be understood apart from our own place in it, for all of us spectators are actors as well, not in the casual way seen in such a play as "The Knight of the Burning Pestle," but occupying the chief parts. Then we confuse the issue by projecting upon the drama mistaken ideas and preconceived versions of the aim of the story. Sometimes these give a false simplicity to the plot, as, for instance, the design-arguments of the Bridgewater Treatises. We must face the facts, and, in forming our theories (this we cannot help doing) we must be prepared to scrap them or revise them, or find them strengthened as the case may be.

What, then, do we see? We see the inter-relation of organism and environment. We shall endeavour to look at the drama of life from each of these standpoints. We do well at the outset of our discussion to remind ourselves that we know nothing of life apart from the organism. We sometimes are apt to speak as though we could see " life " as an entity, as a fact observable by itself, when we know quite well that we can only see it manifested in living forms. Then we cannot study the organism apart from its environment, we sometimes think we can, but it becomes the study of mummies and hay, and has been a fruitful source of mistaken ideas. It is also well to remind ourselves not to give too restricted a meaning to the term " environment." Under this term we include all the factors which have influenced or can influence the organism.

Livingness, then, is the interplay of organism and environment, and the great Darwinian conception of the " web of life " is thus early forced upon our notice.

The drama of life is this interplay, at first little more than a nutritive nexus, but, even in the case of the simplest organisms known to us, there is something more. Perhaps no one has kept this interplay of organism and environment so well to the fore in all his biological studies as Professor J. Arthur Thomson, of Aberdeen. All biologists, indeed all students and lovers of Nature, are indebted to Professor Thomson, not only for his contribution to the study of life, but for the masterly way in which he deals with the problems of biology. He uses words with the skill of an artist, and his descriptions are beautiful and vivid, and yet simple and true like all good art. He is above all an interpreter, who, with wide knowledge and deep sympathies and wonderful skill in the language of explanations, paints his unbiassed pictures. He enables us to see the life of organisms and to see it in all its bearings ; and yet he leaves us in no doubt that to him there is a meaning and a purpose.

As we contemplate the drama of life among plants and animals, both as we can see it around us with our eyes, and as we can see it with the help of telephotic apparatus (such as the microscope and the palæontological museum !), we discern one perennial problem and endeavour, namely, to adjust relations between the active, self-assertive, insistent, insurgent organism and the environment. The inorganic environment is callous, irresponsive, heavy-handed, yet remarkably amenable to life's purposes; the organic environment is capricious, unpredictable, combative. On the one hand, we see the Environment acting upon the organism, burning it and stoking it, heating it and cooling it, quickening it and slowing it, moistening it and drying it, provoking it and quieting it, nurturing it and killing it, cradling it and burying it. On the other hand, we see the Organism responding to the environment, operating on it, changing it ; thrusting as well as parrying ; defying it, mastering it, and using it ; even selecting it. Now the business of life is the continual adjustment of this two-fold relation. But when we look more closely into the effective, regulated, self-assertive, self-expressive, insurgent activity which we call "life," we see that it takes two main directions—caring for self and caring for others. That is the two-fold business of life which we all pursue,—the half-awake plant, the dreamy coral, the instinctive ant, the intelligent beaver, and rational man. The imperious primal impulses are 'Hunger' and 'Love,' the subject and counter-subject of the great fugue of life.[1]

As the play proceeds other things become evident. Although the organisms exhibit what

[1] *The System of Animate Nature*, p. 291.

we may call the fact of heredity—like reproducing like—there is present the equally insistent fact of variation. The first provides that there is a real continuity from parent to offspring, a continuity that is a necessary condition for order; the second provides the raw material of change, which by the principle of inheritance is not lost or swamped, but handed on to the offspring. There is no need to spend time here in giving the evidences of variation, there are many chapters in many books devoted to the subject, and we are all conscious of this fact of variation, even if our observations go no farther than the members of our own family, or those round about us. It is the fact of variation that concerns us here, not the interesting details of the various theories about it. Whether we consider the slight disturbances from the normal, or the big disturbances, to which the name "mutation" is given, it is clear that they occur, and that somehow or other they arise in the germ-cells and show themselves as development proceeds. These variations are like so many experiments made by man when facing the problems in his life, and they are tested in much the same way by their success or failure.

Is there anything in the life of organisms which tests these variation-experiments? Biologists recognise "Natural Selection," which can be translated as "the sifting power of Nature." Man's experiments, inventions of his thought, are subjected to the pragmatic test, "do they work?" The gardener selects for propagation those species which show a variation he wishes to perpetuate.

The breeder does the same. In the world of Nature animals and plants exhibit their strange variation-experiments, and " Natural Selection " decides whether they are to survive or not. It is the pragmatic test. Here in a sentence is a definition of the process.

> Natural Selection may be described as the process by which, in the struggle for existence, certain variants of a species, marked from their fellows by the presence or absence of some innate character, are on that very account favoured with longer life or with more successful families than their neighbours, who are on that account sooner or later eliminated.

The two conditions necessary, then, for Natural Selection to take place, are variation and the struggle for existence. In other words, the organism and the environment. There is the wonderful creative power of the organism shown, not only in these large or small variations, but shown in all its varied response to the stimuli and pressures of the environment. If we also remember the " Web of Life " idea and all that this means in the action and reaction between organisms and their environment, inorganic as well as organic, we shall agree that the Darwinian theory of selection has much to commend it as an explanation of the survival value of organisms and of their parts.

This account of the matter, brief as it is, may seem to many to make out a case for a system which is too mechanical, and which is open to too many objections. This criticism is just if we take for granted the ordinary connotation of the term "struggle for existence," and if we argue from the very restricted idea of what an organism is. We have already given some idea of what, in the light of recent research, we understand by an organism. Let us summarise these in order to have them in mind as we consider the larger and more difficult problem of the reactions of organisms to their limitations and difficulties, which is a more euphonious as well as a more accurate way of describing the struggle for life.

Personality persists in spite of a ceaseless metabolism or change. The animal builds up and breaks down ceaselessly, and yet its individuality remains the same. We sometimes say that every part of the body of a man is changed in a cycle of seven years, and yet he is the same personality throughout. The whole question is really much more subtle than this, and has been well described in the following paragraph as

> a constant warfare between the katabolic and anabolic principles, the katabolic prevailing always it is true—death always wins the victory over the individual—but at each successive stage the anabolic principle making a better bid for victory by forming more efficient energy-stores for the purpose of controlling more effectively the animal's destiny.

The organism, by reason of this storing, can grow; the growth taking place by division and multiplication of cells; and the question of reproduction as well as other kinds of specialised growth have been illuminated for us by Driesch and other experimental embryologists. But there is something more in a living animal than these things, important and distinctive as they are, and this something more is the most important feature of its make-up from the point of view of our enquiry. It may be expressed by the term "creative individuality."

All living animals exhibit what biologists call "behaviour," that is to say, they are agents; their actions are not fixed, they depend upon the peculiar circumstances of the moment. The response they make cannot be decided beforehand, it is contingent. Even among the protozoa, the very simplest of all animals, something like behaviour has been observed.

Another feature is the breaking of fresh ground, made possible by reason of the variability of the organism; and these evidences of creative power are the more important and valuable because the experience gained by this special behaviour can be registered or remembered for a time, it may be for a long time and in some cases for all time, in which case the variation-experiments become part and parcel of the organism.

These, then, are the characters which belong to the living organism. Now, what of the "struggle for existence" in which this organism is supposed to be immersed, more animals are born than can possibly survive. There is therefore a competition or struggle for life, and the fittest survive. But the matter is not so simple as it seems. A superficial view of the facts gives a one-sided impression, and a one-sided impression has, unfortunately, received the sanction of biologists of note. The fact that Huxley gave support to this incomplete view is singularly unfortunate, for his fine publicity work and his bold championing of Darwinism against the bigotry and prejudice of the day, caused his views to be widely known, and these views were accepted as typical of all biologists. It has been one of the hardest of tasks to get a truer, a more balanced view, a view for which Darwin himself pleaded, accepted by the ordinary educated man. Darwin in *The Origin of Species* pleaded for a wide and metaphorical meaning to be given to the term "the struggle for existence." He urged that it must cover success in leaving progeny and the interdependence of organisms. To call the drama a "huge gladiatorial show" is not to take cognisance of all the facts.

If we think of the struggle as the reaction of the creature against its difficulties and limitations, we shall see that there are three of these difficulties, (1) over-population and the scramble for food, (2) the vegetarian and the carnivore, (3) the changing physical environment. All these difficulties belong to the environment of the organism, and the "struggle" is the general term for the way in which the creative self-assertiveness of the organism deals with these difficulties. Every effective adaptation which survives will bring the creature into a wider environment, and at the same time will being fresh difficulties and fresh problems into the lives of other organisms. The web of life is a true picture. We must not stay now to continue this thought about the widening of the environmental stimuli by the struggles of the organism; we shall revert to it a little later. We have already referred in this chapter to the conception of the life of the organism which Professor Thomson gives, and we have recognised the two-fold business of life in the care for self on the one hand and the care for others on the other. This latter business involves sacrifice and subordination of personal ends for the sake of offspring. This side of the story is often forgotten, and must be placed alongside the story of gladiatorial display. But even in the other business of life, the care of self, selfishness is not the last word.

Professor Henry Drummond[1] has helped many to correct their crude ideas, and Kropotkin, in his essays on *Mutual Aid*, brings all the weight of his biological and sociological knowledge to the task of correcting the one-sided views of Huxley. Kropotkin shows very clearly by many examples that combination or sociability is one of the replies which organisms make to the limitations of the environment. It will be well to quote a paragraph.

That life in societies is the most powerful weapon in the struggle for life, taken in its widest sense, has been illustrated by several examples on the foregoing pages, and could be illustrated by any amount of evidence, if further evidence were required. Life in societies enables the feeblest insects, the feeblest birds, and the feeblest mammals to resist, or to protect themselves from, the

[1] *The Ascent of Man*, chapter vii.

most terrible birds and beasts of prey ; it permits longevity ; it enables the species to rear its progeny with the least waste of energy and to maintain its numbers albeit a very slow birth-rate ; it enables the gregarious animals to migrate in search of new abodes. Therefore, while fully admitting that force, swiftness, protective colours, cunningness, and endurance to hunger and cold, which are mentioned by Darwin and Wallace, are so many qualities making the individual, or the species, the fittest under certain circumstances, we maintain that under *any* circumstances sociability is the greatest advantage in the struggle for life. Those species which willingly or unwillingly abandon it are doomed to decay ; while those animals which know best how to combine, have the greatest chances of survival and of further evolution, although they may be inferior to others in *each* of the faculties enumerated by Darwin and Wallace, save the intellectual faculty. The highest vertebrates, and especially mankind, are the best proof of this assertion. As to the intellectual faculty, while every Darwinist will agree with Darwin that it is the most powerful arm in the struggle for life, and the most powerful factor of further evolution, he also will admit that intelligence is an eminently social faculty. Language, imitation, and accumulated experience are so many elements of growing intelligence of which the unsociable animal is deprived. Therefore we find, at the top of each class of animals, the ants, the parrots, and the monkeys, all combining the greatest sociability with the highest development of intelligence. The fittest are thus the most sociable animals, and sociability appears as the chief factor of evolution, both directly, by securing the well-being of the species while diminishing the waste of energy, and indirectly, by favouring the growth of intelligence.[1]

As a footnote to this the story of Silverspot can be read in *Wild Animals I have Known.*

[1] *Mutual Aid,* 1904, p. 57.

The consideration of this aspect of the problem has been carried a stage further by Trotter in his *Instincts of the Herd in Peace and War.*[1] His ideas were put forward some years before the war in the pages of the *Sociological Review,* but as the war brought to a head the question of the struggle for existence in a vital way, and caused a general review of the biological background of human society, so this study of herd instincts had its thesis tested by the behaviour of man during war. The author was able to develop his original essays into one of the most important of modern biological writings.

Let us consider the general problem of gregariousness in the lower animals, leaving for the moment the special case of man.

In an interesting argument, Trotter discusses the change that comes over the life of a single-celled animal in the change to a multicellular one, looking upon the multicellular one as in the first instance, at any rate, an aggregate of equally efficient cells. He shows that as a result of association the individual cells, while losing some of their independence of action, are withdrawn to a considerable extent from the severities of the struggle for existence and that thereby there is more chance for the persistence of all manner of variations.

> It would seem, therefore, that there is now allowed a greater range of variability for the individual cells, and perhaps, therefore, an increased richness of the material to be selected from. Variations, moreover, which were

[1] Published in 1916.

not immediately favourable would now have a chance of surviving.[1]

In a similar way the rise into gregariousness is a method of meeting the limitations and difficulties of the environment, and at the same time of allowing freedom to vary and to specialise in safety. In surveying the behaviour of social animals certain principles stand out clearly. The herd or group acts in association for offensive or defensive tactics, the majority are defensive herds. Within the herd there must be a sensitiveness to the behaviour of others, and a power of communication. The individuals must be open to suggestion. Leaders are necessary, and are followed and obeyed so long as they keep in touch with the herd, and do not go too far ahead. Loneliness is a terror, and the fear of isolation from the herd is very great. In some communities the individual cannot live apart from the herd. There is great freedom for the individual within the herd, and two main types can be distinguished, the normal or stable, and the specialist or instable. These types are referred to under the terms "co-ordination" and "specialisation." There are those who lean towards co-ordination, and there are those who are keen upon specialisation. Of course we cannot find a perfect dichotomy any more than we agree with the sentry in *Iolanthe*—

> That ev'ry boy and ev'ry gal,
> That's born into the world alive,
> Is either a little Liberal,
> Or else a little Conservative!

[1] *Trotter*, p. 18.

Still, in general, these two types are found, and both are necessary to the success of the community, and probably all individuals exhibit both characteristics in some measure. The visions and experiments of the specialists must be controlled and co-ordinated with the main ideas of the herd; but the suppression of specialisation and the loss of experiments will bring about stagnation and extinction in the long run. In looking at a number of animal communities we can see how these two characteristics play their part. In the beehive or the anthill we see co-ordination in its extreme form. Specialisation to a bee or an ant is extremely limited, and this is clearly related to the small brain and to its consequent instinctive habits. In the life of rooks or parrots, beavers or the social monkeys, the greater brain power makes possible a wider response to the limitations and difficulties of the environment. In these communities of more intelligent animals there is a greater balance between specialisation and co-ordination. The character of the herd or community will have its effect upon the minds of the members of the herd, and will determine in large measure the normal behaviour. Man is a gregarious animal, and the whole question of the effects of this gregariousness upon his behaviour and upon his mind must be taken into account in relation to his evolution.

We must be careful not to argue from a community to an individual, nor from an individual to a community, for, especially in the case of man, the individual as an individual retains a clear autonomy. However, at the moment we are not concerned with pressing the argument further; what we do need to realise is that the ascent of man has been the resultant of many pressures, that the way of association has been the main answer of man to these limitations and difficulties, and that in recognising sociability as a factor in evolution we also recognise its results in fostering intelligence and experimentation.

* * * *

Let us now attempt to see this concept of evolution from a different angle. We have been trying to get some idea of the organism, with its self-assertiveness and its experiment-variations, facing the difficulties in its environment. The responses made in face of these difficulties are various. Some are ignoble responses, such as parasitism; there may be degeneration or stagnation as a result of sedentariness; but, taking a general view of the whole story, it is clear that there has been an increase in complexity and in correlation, or, in other words, there has been progress. We can see progress in structure and progress in adaptation to an ever-increasing environment. That is to say, ever increasing from the animal's point of view, for, as we shall see, the fuller environment is present before the organism becomes adapted to it. In a very real sense we can say that a reptile is higher than a fish, and a man is higher than a reptile. We do not say this because we wish to flatter man, but because we can see that man has learnt to control and use a bigger environment than a reptile. Man's life has a bigger contact with the environment. Man has a more complex behaviour, for his mind has come into conscious operation and takes its place in the ordering of his life.

So, as we look upon the whole drama of the evolutionary story from this angle, we see a wonderful orderly system, with ever-increasing complexity, and ever-increasing adaptability, along with increasing knowledge and use of the environment of natural law, for regularity is the characteristic of the inorganic domain.

Watch an amœba on the microscope stage. It has but feeble powers of appreciating and responding to the stimuli around. It has an inherent irritability, but of a very diffuse and general kind. It moves slowly across the field of vision, and we can see it accept or refuse the diatom lurking in its path. As Jennings[1] has shown by some interesting experiments even an amœba has the beginnings of behaviour. Watch a fish swimming freely in the pond or river. It is aware of much more in the environment than the amœba sliding over the mud at the bottom of the pond. It has an elaborate eye for registering light waves, an ear for registering sound waves, elaborate sense organs for smell, for testing the pressure of the water, and for balancing itself. It has complicated organs for respiration and for dealing with its food, necessary for supplying the needs of a complex body with its great development of muscles, which are largely used for

[1] *Behaviour of the Lower Organisms.*

locomotion in the water. Its behaviour is correspondingly more elaborate. The formation of an eye—and between the amœba with its general sensitiveness to light and the fish there are many experiments in eye-structures—is meaningless without an environment in which it can be used. Fins and a muscular body ending in a tail are meaningless except as we see them used for locomotion in a medium suitable for their effective use.

Watch the rabbit feeding at the side of the river and compare its behaviour with that of the fish. An entirely new set of environmental stimuli play upon it. Its life is adapted to the dry ground, with a new method of locomotion and changed musculation operating a changed framework; it breathes the air and its sense organs are adapted for use in the air. Its behaviour is more complex than that of the fish, it has consciousness and an intelligence more subtle. And now add to our picture a man walking across the meadow near by. He is conscious of himself, and guides his life by a combination of intuition and thought. He is thrilled by the beauty of the landscape, and finds some curious inward satisfaction in following the call of goodness and truth. He is certainly living in a higher or fuller environment than the rabbit, and the fact that his personality can be stirred by beauty is as much a demonstration of the reality of beauty in the environment, as the possession of eyes demonstrates the reality of light waves. The argument is briefly this. Just as the environment to which the fish responds was there for the amœba although all unknown and unused, just as the air and land environment has been there even though the fish is not conscious of it, so the spiritual environment (the environment of which beauty is a part) to which man responds has been there all the time waiting for the recognition and use by man. What we have been seeing in the drama of life has been the gradual appreciation of fresh factors of the complete environment, and this gradual appreciation has been possible by the creative experimentation of the organisms. It looks as though a fresh environment is being formed at every big stage in the story, like a change of scene at the theatre, but a moment's thought will convince us that the higher environment must have been there all the time, otherwise the variations towards fins or towards lungs could never have been fixed, and could never have been gradually developed into their present wonderful structure. Thus we have got to this place in our thought, that the perfect or full environment has been present from the beginning, and indeed, if our argument is sound, even before the first great beginning of life in the primordial organisms. There had to be a preparation of the world before life was possible, and geologists and astronomers have much to tell us of the early stages in the genesis of the world. This preparation of the inorganic world to be part of the environment or cradle for the organic is surely part of the scheme as we are attempting to visualise it. The environment necessary for life must have been present around that mysterious chemical bench from whence it seems that organisms took their initial step into the unknown. We are speaking of course in metaphors, and much is speculative, and we do well to remember that continuity does not only mean the rise of man as the natural and inevitable working out of the scheme, but that at the other end of the scale it means the continuity in some unknown way between the organic and the inorganic. We divide the world by three cross-cuts; we do it for clearness, and there is much to commend it from this point of view; we speak of the non-living, the living organism, and man the spiritual being. It is certain that wide gaps seem to divide these three groups, look at them how we will; but yet we feel that the truth is rather in this view of continuity. There have been great inventions—life itself, insects, birds, man—yet these inventions, at least those resulting in insects, birds, and man, can be understood in the light of this interaction of organism and environment. The gaps in our story are measures of our ignorance, and there has been considerable bridging of gaps during the last fifty years, as our knowledge and insight have increased. It may be that the gap between the organic and inorganic will be bridged sooner or later, and this need not trouble us or upset us; indeed the idea of continuity is bound up with our vision of the all-pervading environment, which did not come into existence when life first appeared on this earth. And at the other end of the drama, as we see the outcome in the personality of man, we feel convinced that personality must be a part of this all-pervading environment. Can this be other than a manifestation of God ?

Some people seem to think that only if unbridgable gaps exist can they believe in God ; they try to justify their belief in God by postulating the need for some power to jump over these gaps. The view indicated here is the opposite one, that in continuity one has a glimpse of a Divine method, orderly and consistent.

The drama of evolution begins then to have something of a plot, and we shall attempt the task of looking for the purpose and meaning of its wonderful age-long process.

BOOKS

GEDDES, P., and J. ARTHUR THOMSON . . . Evolution.

SCOTT, W. B. . . . The Theory of Evolution (New York).

THOMSON, J. ARTHUR . . . The Study of Animal Life
The System of Animate Nature.

(The first of these books has a useful bibliography for the general reader.)

CHAPTER IV

THE PURPOSE OF EVOLUTION

> " I heard it all, each, every note
> Of every lung and tongue and throat,
> Ay, every rhythm and rhyme
> Of everything that lives and loves,
> And upward, ever upward, moves
> From lowly to sublime ! "
>
> RALPH HODGSON.

THE subject we attempted to consider in the last chapter was the drama of life, but our consideration was too slight to convey anything approaching the grandeur of the story. The more one studies it, the more profound and complex does it become. If this is true of the concept of evolution, how inadequate must be our treatment of the problem before us now. The most we can hope to accomplish is to help one another to get a clearer idea of the kind of answer that can be given to the " Why ? " and " Wherefore ? " The answer, however incomplete, must be the outcome of the consideration of the facts of evolution.

What is the purpose of evolution ? What is it all for ? Granted that the method of evolution is the only explanation of the facts, why was that method chosen ? Can we detect any purpose, any aim, which explains the age-long process? Although one is open to the sneer of the sceptic that an outsider might have a different opinion, yet the story does seem to find its explanation and its purpose in the advent of man as a personality. It does seem as though the answer to our question will be found in the spiritual consciousness of man as realised in fellowship with God. The fact that man is able to ask these questions, and, in some feeble way, is able to comprehend the purpose encourage us to go forward; and his attempts to identify himself with that purpose and thereby to further it give us the hope that an answer will be found along the pathway of fellowship and service.

We need not mind if, in attempting to find the answer, we go beyond the limits of science, for any satisfying answer must be in a language that transcends the language of science. Religion brings another contribution towards the answer, and this contribution working downwards from man's personal experience of God should find its confirmation in the contribution from the scientific aspect which, if we may use the metaphor, works upward to meet it. Religion surely deals with the struggles of man's personality amid the limitations and difficulties of his environment, and any explanation we may hope to give cannot really satisfy unless it receives support from religion. The one hope that is with me throughout this enquiry is that the labours of scientists may make the answer, from the religious side, more reasonable and acceptable. The work of the scientists should provide a highway leading to a place of vision, from which the spiritual faculties enable one to catch sight of the city of God.

We watch an artist busy at his canvas endeavouring to represent some idea he has in his mind; we have no right to expect to read his message until the picture is nearing its completion. A schoolmaster with new ideas like Job Huss, in Well's *The Undying Fire*, is open to destructive criticism until he can show the result in the kind of men turned out by the school and in their attitude to the problems of life. So the scheme of evolution must not be judged by this or that element in the composition of the picture taken apart from the whole idea; rather we must judge it by the results. At the same time, it must be confessed that the picture has many disfigurements which have been caused by man's ignorance and by his wilful disregard of the purpose; these disfigurements must not be accredited to the scheme.

But is there a scheme?

I believe a careful study of the facts will lead us to discern a directive power working from within, which power, at the lower end, seems to be concerned with the more mechanical shaping of passive material for use, although this is never entirely mechanical; and, at the upper end, in conscious relation with God, can direct our lives to good purpose. In spite of disharmonies and discords, in spite of difficulties and obscurities, there is a melody to be heard, and if a melody surely there is the artist-mind that planned it—there is the Musician,

The inorganic domain with its stable laws and orderly happenings provides the setting for the rise of individuals with their adaptations and their gradually increasing mentality. The idea of design and designer is not less wonderful because we can trace the beginnings far back, and can see the working out along the lines of the interaction of organism and environment—surely it is more wonderful. Creative mind is more applicable to an orderly, continuous scheme than to a scheme requiring constant interference and fresh creations to help it on its way.

But why this method ? The answer can be given in four words—the winning of freedom.

There is no need to discuss here the characteristics of personality. It has been discussed and argued about from every conceivable angle, and the fact of man's freedom is acknowledged by all. We can follow Illingworth's description as a moderate statement, and see in the development of man's personality a later chapter in the evolutionary story. Man comes to himself, and the revelation by the way of self-consciousness leads to the knowledge of his freedom. He knows that he can choose his way. This is not absolute freedom, it means freedom relative to the conditions in which he finds himself. The way man has come is a very definite pathway. His steps in the past have conditioned, to a certain extent, the possibilities of the present, and every step forward conditions, in some measure, the next possibility.

But man finds himself free in this, that he can within these limits choose his way, and that as he begins to realise the factors which play about his path, he finds the choice a wider one. In exercising this power of choice man finds himself in a bigger environment, and finds subtler forces playing about his path leading him onward into a life which begins to exhibit a purpose and a plan. The purpose makes the past story understandable.

How has this freedom been won? By struggle and endeavour with conditions which are not free. The whole story of evolution, the struggle —action and reaction—between organism and environment is really the evolution of freedom, and thus of personality. There is no other way in which freedom can be obtained; it cannot be given, it can only be won. It has to be won right from the beginning, otherwise it would not be real freedom. Thus our picture of the creative power pushing out into the unknown and untried, even in the simplest organisms, and in those complex-simple microcosms the germ-cells, are really illustrations of the same story as that seen in the creative power of man. This has been possible because of that.

The facts of biology help us in our argument. Even in the amœba there has been seen something of behaviour, some primitive freedom in the choice of the paths along which to go. Of course the choice in this lowly organism is not a conscious one; it is the fact that there is something of a choice seen in these lowest organisms that is of importance. We can trace the rise from this slight beginning up into the state of a conscious choice, as the organism increases in complexity and has an increasing realisation of fresh parts of its environment.

The purpose of evolution is man,—but what man ? The prehistoric ape-man or the man that is to be ? The end is surely not in Homo animans, nor in Homo faber, nor even in Homo sapiens, although each of these marks a mighty step along the way. Is it in Homo spiritus ? We may be allowed to suggest that this is not the end, but that, looking still ahead, we see in Homo deus, man as a son of God, a fellow-heir with Jesus and fellow-worker with God, the aim and purpose of it all. But this you will say is going too far ahead of our facts. Can we demonstrate in any way that these powers and characteristics, which we recognise in human personality, are the outcome of the scheme ? Has Biology any evidence to offer as to the spiritual nature of man which supports this momentous claim ? Is Homo spiritus the descendant of Homo animans ? I believe there are facts to bring forward which help, in some measure, to make the claim a real one ?

We have passed rather lightly over the later stages of man's evolution in its phylogenetic phase. In Chapter I we mentioned the upright carriage freeing the hand for use as an instrument in tool-making, and a large brain, as aids to his progress, and also referred to the fact that life in societies was probably an important factor. Many consider this last as the most important factor, even as the real factor, and regard the others as subsidiary; sociality providing the freedom for the development of the other factors. In any attempt to understand the rise into a moral and spiritual person, the importance of the gregarious factor is still further emphasised. In meeting the limitations and difficulties in the environment, gregariousness has always been a favourable method, and one much more widespread than most people realise. It is opposite in its action to the crude "fight for life," for it allows freedom and scope for the development of the finer qualities of altruism and sympathy and favours intelligence. We have here a clue to the factors of man's ascent, and the statement that "society made man and not man society" is certainly true if we can put it in this form—"society has made it possible for man to make himself."

It will be necessary to enquire rather more fully into this important aspect of the subject, for considerable attention is being directed to it from several quarters, from the behaviourists as well as from the psychological and sociological sides. The war provided a test for rival theories of "conscience" and "herd behaviour," and many ideas have been reviewed in the fierce light of this experience.

We have already enumerated the main characteristics of gregarious animals, and it will be an easy matter to establish the fact of man's gregariousness, although the effect of this habit upon his mental and moral make-up may not be so readily recognised. Trotter's treatment of this subject in his *Instincts of the Herd in Peace and War* is interesting and important, but we shall not follow him in many of his conclusions.

It is necessary to bear in mind the value of sociality as an evolutionary method, and to realise how widespread have been the movements towards interdependence. As we have said above, association is one of the effective ways whereby organisms can meet and conquer their limitations and difficulties. By it they enlarge their sphere of activity, and increase their hold over matter, and bend more and more of the environment into their service. This is largely the result of the relief gained by association from the cruder pressures of the struggle for existence, for by this relief and shelter individuals of the group are more free to exhibit signs of specialisation. In more biological language association seems to favour the production of variations, and to afford these variations, even if not directly favourable, a better chance of survival. If we remember that variations are the raw material of progress, and that, by being continued in the germ-plasm, they count repeatedly in the struggle for life, we shall realise the value of association in evolution-story. There is more in the social life of man than this sheltering and freedom for variation-experiments, important as these have been, especially those related to the changes in the carriage of the body, the use of the hand, and the development of greater intelligence. There is such a thing as society-variation, when a whole group may react, as a group, to its difficulties in a change of environment. Such a reaction is seen in the change from tree to ground life, or in the migration from slavery in Egypt to a nomadic life in the desert.

There is also another kind of influence which comes into play as sociality develops into a permanent habit. This is sometimes called "social heredity" to distinguish it from "individual heredity," but perhaps the term "social heritage" is a better term. It means that from birth the new individual is subjected to influences and repressions, customs and taboos, which are largely responsible for the way in which the development of the individual inheritance proceeds. These influences, to which the general term "nurture" is given, are the environmental forces which allow or prevent the normal development of "nature," the inborn characteristics and variations. Where the development is largely a question of mental and spiritual development, as it is in man, the importance of "nurture" can hardly be over-emphasised. The importance of considering the social habit and its bearing upon the ascent of man will become clearer as we consider its various aspects.

Comparing man with other gregarious animals the main likenesses can be grouped under the following heads—sensitiveness to the voice of the herd, liability to panics and mob passion, susceptibility to leadership, recognition of the other members of the group and distrust of those of another group, intolerance and fear of solitude.

The herd for its success must have solidarity or

unity, and for this the power of communication is necessary, and the corresponding sensitivenes of the individuals to suggestion from others. It is unnecessary to discuss here the important advance which speech has made in the suggestibility powers, or to follow out the further advance when these herd ideas become inshrined in literature, the oral tradition becoming permanent. Every advance carries with it dangers and difficulties; for instance, an oral tradition may show a growth alongside the developing social life which is denied to a code which, by being written, remains too rigid.

It may be well to point out here the caution against a careless passage from the gregariousness of lower animals to that of man. No doubt there are general characteristics common to all associations of organisms, but there are fundamental differences between the associations of men and those of lower animals which may vitiate any argument based upon a common gregariousness. Especially is this so with regard to the associations of ants and bees. In these social groups we find wonderful co-ordination, and the actions of the individual insects almost entirely instinctive, with a curious corporate instinct of the hive or nest. This makes the whole hive to have the character of an individual, and a bee cannot live an isolated life, any more than an organ or a cell of the body. Although we can see a similar tendency in the social life of man, when by an elaborate division of labour individual men are less and less capable of maintaining a solitary existence, and depend upon one another for the outward necessities of life; yet this inter-relation among men is a very different thing from the division of labour among the inhabitants of an ant-hill, or among the cells of the body. The gregariousness of man is a secondary grouping; it has come about in response to special needs and conditions; it is a grouping of already highly intelligent animals. Thus the association together began as a conscious association with a recognition of its value, even if this value was only at first a dim idea of safety and mutual aid. We will return to these differences below.

Let us now consider one or two points which make it clear that the gregarious habit has a profound effect upon the life of man. A little calm reflection will convince us that a great part of our beliefs, which we fondly imagine are founded upon reason, are, in fact, non-rational, and have been enforced upon us as the code of the special class to which we belong. The herd instinct, working through education and utilising convention, good form, and traditions, provides the mechanism whereby this enforcement is accomplished. The following extract from Trotter is worth quoting:

Direct observation of man reveals at once the fact that a very considerable proportion of his beliefs are non-rational to a degree which is immediately obvious without any special examination, and with no special resources other than common knowledge. If we examine the mental furniture of the average man, we shall find it made up of a vast number of judgments of a very precise kind upon subjects of very great variety, complexity, and difficulty. He will have fairly settled views upon the origin and nature of the universe, and upon what he will probably call its meaning; he will have conclusions as to what is to happen to him at death and after, as to what is and what should be the basis of conduct. He will know how the country should be governed, and why it is going to the dogs, why this piece of legislation is good and that bad. He will have strong views upon military and naval strategy, the principles of taxation, the use of alcohol and vaccination, the treatment of influenza, the prevention of hydrophobia, upon municipal trading, the teaching of Greek, upon what is permissible in art, satisfactory in literature, and hopeful in science.

The bulk of such opinions must necessarily be without rational basis, since many of them are concerned with problems admitted by the expert to be still unsolved, while as to the rest it is clear that the training and experience of no average man can qualify him to have any opinion upon them at all.[1]

Besides this sensitiveness to the voice of the herd, and the fact that many opinions are held simply in virtue of our group life, and, as such, held with surprising firmness, we are aware, unhappily, of the ease with which we fall into sub-groups and classes, and how difficult it is to get real sympathy and true understanding between the individuals of different sub-groups. This is largely the result of excessive herd suggestion, and illustrates how little reason controls our actions and opinions. Then the facts of mob violence and crowd enthusiasm are known only too well. A mob swayed by anger or emotion will sanction and perform actions which in its calmer moments it would repudiate, whether the excesses attend a lynching or a revival. This is

[1] *Instincts of the Herd in Peace and War*, pp. 35 and 36.

a case of the more primitive herd, easily swayed, dangerous, and irrational; fortunately it plays a lessening part in the more complex social life of man, but during the war there were many ugly happenings showing still how easily it may be excited.

Another characteristic usually attributed to social life is the intolerance of solitude. It is certainly true that man finds his deepest satisfaction in fellowship. There is a feeling of incompleteness, and this is only met by the freer and higher life found in association with others. Trotter traces this feeling of incompleteness, and this desire for union in a bigger life up into the religious ideas of spiritual fellowship. To trace the development of morality in relation to social life does not explain its origin, nor does the fact that social life gives the chance for the development of the spiritual nature of man prove that it originated in this way. Perhaps it will be well to let Trotter state the case as he sees it.

It is obvious that when free communication is possible by speech, the expressed approval or disapproval of the herd will acquire the qualities of identity or dissociation from the herd respectively. To know that he is doing what would arouse the disapproval of the herd will bring to the individual the same profound sense of discomfort which would accompany actual physical separation, while to know that he is doing what the herd would approve will give him a sense of rightness, of gusto, and of stimulus which would accompany physical presence in the herd and response to its mandates. In both cases it is clear that no actual expression by the herd is necessary to arouse the appropriate feelings, which would come from within and have, in fact, the qualities which are recognised in the dictates of conscience. Conscience, then, and the feelings of guilt and of duty are the peculiar possessions of the gregarious animal.[1]

A few pages further on he writes:

Religion has always been to man an intensely serious matter, and when we realise its biological significance we can see that this is due to a deeply ingrained need of his mind. The individual of a gregarious species can never be truly independent and self-sufficient. Natural selection has ensured that as an individual he must have an abiding sense of incompleteness, which, as thought develops in complexity, will come to be more and more abstractly expressed. This is the psychological germ which expresses itself in the religious feelings, in the desire for completion, for mystical union, for incorporation with the infinite, which are all provided for in Christianity and in all the successful sub-varieties of Christianity which modern times have seen develop.[2]

I have given these quotations from Trotter because they illustrate the position taken by a large number of biologists, especially those interested in sociology, and they seem to press what is an important point of view to an extreme position, confusing development with origin. The view taken here, as outlined in an earlier chapter, is that, agreeing with all that is urged about the relation of morality to racial good and herd requirements, realising that social life has brought into great prominence the conflict between the general good and the individual satisfactions,

[1] *Instincts of the Herd in Peace and War*, p. 40.
[2] Ibid., p. 50.

and that under the term " conscience " is included much that is an insistence of racial experience, the conflict did not originate during the period of man's gregariousness, but belongs to the whole of life. There is somewhat in man able to react under the stimulus of sociality; the conditions and circumstances of social life give a great chance to these inward characteristics to develop. There is somewhat in man which is the source of spiritual life; this is not given to him by any outward organisation of herd or community, important as these are in its development, but it is a part of his personality, and as such is naturally concerned with communion and fellowship. I believe that a fuller study of the gregarious habit in man will support this view. We have already referred to the important features which distinguish the social habit in man, its late adoption and the importance of intelligence. We do not know at what stage in his phylogeny the social habit became permanent. Probably it was in those early days, when as a simple Primate, life on the ground, with upright carriage and a freed hand, was adopted. Gregariousness in this case became a very important " nurtural " pressure, and the dictates of the herd would come into conflict with the well-established individual primitive instincts. This nurtural pressure has increased and developed along with the development of man's power of taking advantage of association, and because of this its effect will be great and deeply ingrained; but we must remember that we are dealing with the conscious association of intelligent and reasoning animals, and that the development of the social life of man has been largely the result of specialisation and experimentation on the part of individuals. The way of the reformer is always against the general herd idea in the first instance, and this strengthens the view that the gregarious habit does not originate the religious instinct.

Group-actions may often be the result of the corporate instinct, but always against this there is the intelligence of the individual, for in the communities of man there is no such thing as a super or corporate intelligence, any more than there is a corporate personality. So in the winning of freedom, although the herd suggestions and prohibitions have had a profound effect, the central problem is an individual matter. The recognition of a "good" and an "end" is always an individual concern, and is the great characteristic of personality. The psychologist in studying the mental conflicts finds that many of them arise by the lower primitive self-instincts, or the religious intuitions of the individual coming into opposition with the line of conduct sanctioned by the herd, and the way out from the conflict is by a modification of the conditions, so that the whole personality can develop. Religion is not a kind of refined herd convention as Trotter seems to suggest, but it belongs to the fundamental make-up of the personality, and as such is often the cause of the individual standing out against the herd suggestion, pointing the way, indeed leading the way, to a better life. However, in a later essay in Trotter's *Instincts of the Herd* the following paragraph is found:

> It is therefore necessary to lay down with the strongest emphasis the proposition that the religious needs and feelings of man are a direct and necessary manifestation of the inheritance of instinct with which he is born, and therefore deserve consideration as respectful and observation as minute as any other biological phenomenon.[1]

In giving a general agreement with this statement it will be necessary to enquire what is meant by instinct, and to consider the part played by them in the life of man. If Trotter means that religious needs and feelings are a manifestation of the herd instinct, I do not agree; but if he means that among the primary instincts of man is found some all-important constituent which is the final authority in conduct and a necessary factor to moral freedom, I do agree. This is not the place to discuss fully the psychological aspect, nor can I hope to do more than give a very superficial survey of the subject as it appeals to me; but something must be said in order to bring the rise into a spiritual being into line with the story of evolution and to see in this the purpose of the whole drama. If we use Bergson's definition of instinct as an intuitive comprehension of life and its needs, we shall see that this is one of the two activities into which the rudimentary mind-consciousness of the simplest animals has developed. This instinctive activity is that part of consciousness by which actions are performed right off

[1] Page 114.

without the need for learning or practice, and is part of the innateness of the animal. The other activity is the intelligent activity, whereby the action is based upon experience and memory. The first group of actions are automatic, and deal with the inward and outward activities necessary to life, whilst the second are concerned with the relation of the animal to the limitations and changeful conditions of its environment and are not automatic but intelligent and purposeful.

In man somewhat the same combination is seen. The intelligent activity has become highly developed into reason, but there are also the primitive instincts which still play a great part in life. These are concerned with nutrition, with self-preservation, with reproduction and the care of offspring, and because of the gregarious habit of man there is the herd instinct also. Man is, as we have seen, a social animal, and this social aspect of his personality moulds and alters many of his simple emotions and actions. But the story of man is special and peculiar in this, that in him the intuitive comprehension of life and its needs has taken a higher form. This is because he has climbed up into a higher environment in which intellectual powers and spiritual faculties are able to function.

A little faithful examination of our own actions will help us to see that they are often based upon intuition rather than on reason. "We know ——," and any enquiry as to "How ?" or "Why ?" is considered unnecessary and even impious. It is true that many of these intuitions are related to the herd, to our membership of a class, or to the obligations imposed upon us by our circle, but this does not exhaust the intuitive perceptions. McDougall has a suggestive sentence, in which he describes the highest stage of conduct as that "in which conduct is regulated by an ideal of conduct that enables a man to act in the way that seems to him right regardless of the praise or blame of his immediate social environment," [1] and Dr. Chalmers Mitchell in his *Evolution and the War*, states the problem in a different way:

> I assert as a biological fact that the moral law is as real and as external to man as the starry vault. It has no secure seat in any single man or in any single nation. It is the work of the blood and tears of long generations of men. It is not in man, inborn or innate, but is enshrined in his tradition, in his customs, in his literature and his religion. Its creation and sustenance are the crowning glory of man, and his consciousness of it puts him in a high place above the animal world. Men live and die; nations rise and fall; but the struggle of individual lives and of individual nations must be measured not by their immediate needs, but as they tend to the debasement or perfection of man's great achievement.

Let us approach the problem from another side, from that of religious experience. Religious men base their conduct upon what they believe to be the Will of God. How do they decide what is the Will of God? Some depend upon the external authority of Church or Bible. For them the sanctions and traditions of the herd have supreme weight. Others find their authority in a more

[1] *An Introduction to Social Psychology*, p. 181.

immediate and personal way through the intuitions received from God in their own hearts.

The mystics of all ages and in all religion claim that to them has been given a personal experience of God, something immediate, deeper than any reason, something which is akin to feeling and yet it transcends feeling. It cannot be adequately reported, for to be communicated it must be translated into language, and words are inadequate to convey its reality. That is why the mystic intuitions are described in metaphor. The attempt to give a reasonable explanation of this inward awareness of harmony and fellowship in the larger spiritual life usually fails. We are suspicious of these transports because we misjudge the explanations and reports, and these explanations may be contradictory or wrong, although the experience behind may be real and vivid. George Fox, the Quaker, called the inward intuitive power "that of God in every one," and certainly, if we were to bring forward the witness of Christian experience to the guiding power of the inward over the outward, we should agree that by this "inner light" the will of God is known. We are dealing here, I believe, with a "faculty" or "instinct," call it what you will, as real and as open to observation and investigation as any of the animal-like primitive instincts discussed by McDougall. I do not suggest that the intuitions of spiritual relations whereby guidance may be received, are the same kind as Bergson's intuitive comprehensions, for these belong to the lower life and its needs, whereas those belong to the higher life and its needs; but the analogy is suggestive. Nor do I believe that this "spiritual instinct" has been developed from those other instincts. It is something distinct and apart from these, although its operations may be similar. It has come into use when the evolution of man reached the stage when its operations were possible. It has not been created by any sudden action from without, but by a development from within. We cannot pursue this subject at greater length here, for we have set out to consider the problem of the spiritual life of man from the biological side. Although there is much that is obscure and difficult to understand, still there seems every reason to believe that the spiritual nature of man, with its strange inward activity, which we have compared with our intuitive comprehension of life, has had a history, and that it has become functional as the animal nature has become controlled, and the cruder instincts sublimated in the service of the ideal.

If the main idea put forward here, that the purpose of the scheme becomes known in the God-consciousness of man, is accepted, we can go forward with renewed hope and courage to the consideration of difficulties and obscurities.

The kind of answer we hope to be able to give to our "whys" and "wherefores" is suggested by some sentences taken from an essay by Prince Eugène Troubetzkoy, Professor of the Philosophy of Law in the University of Moscow.

Conceiving our life in this manner, the material evolution of the world becomes the incarnation and the expression of a spiritual meaning, of a divine event which is actually in process of coming to pass. No longer, for example, do we think of the earth's movement round the sun as a meaningless rotation: we think of it as preparing the conditions which enable life to rise to its sublimest height; we see the whole creation saturated in sunlight. Not in vain are the heavens starred with innumerable fires. They speak to us of worlds to which they give life and being, warming them with their heat, brightening them with their beams. And the end to which all these lines are moving, of every flower that blooms, of every bird that sings, is also the central principle of the entire evolution of the Universe—the embodied Word of God. For the purpose of the whole is nothing other than the incarnation of the divine, the participation of the created in the eternal life of the uncreated, of which the God-man is the perfect revelation.[1]

BOOKS

McDougall, W. . An Introduction to Social Psychology.
Simpson, J. Y. . The Spiritual Interpretation of Nature.
Tansley, A. G. . The New Psychology.
Trotter, W. . . Instincts of the Herd in Peace and War.

[1] *The Hibbert Journal*, vol. xvi. No. 3.

CHAPTER V

DOUBTS AND DIFFICULTIES

"Roaming in thought over the Universe, I saw the little that is Good steadily hastening towards immortality,
And the vast of all that is call'd Evil I saw hastening to merge itself and become lost and dead."

WALT WHITMAN.

"The world were an utter and hopeless mystery if pain were not. Where, then, would be the basis and the root of love, the prophecy of an enlarged and an ennobled nature? where the revelation of our life in Christ?"

JAMES HINTON.

MOST of us would express the problem of life differently. To some, James Hinton has put it the wrong way round, for to them it is the presence of pain and evil which makes the mystery. Others will not share Whitman's optimism in seeing the hastening victory of good. Both views are, I believe, correct. The meaning of the world's life is understandable only on the basis of the triumph of goodness; and it is a true insight that sees the necessity of pain and evil.

Some may take the line that an answer to the doubts and difficulties is not possible and that man must trust that somehow things will work out all right. God must be trusted; we, who are finite and insignificant, how can we hope to fathom the mind and purposes of God, the Infinite. This attitude is possible for some, but for a decreasing number. Man is always a questioner, he is full of curiosity, it is in his blood, for without it he would have never become man. He will always be asking "Why?" "How?" and "Wherefore?" If he is called upon to trust the larger hope, he wishes to have a reasonable ground for his trust. He wishes to face life to see if it is trustworthy or not. Thoreau faced it beside the Concord Pond.

> I went to the woods because I wished to live deliberately, to front only the essential facts of life, and see if I could not learn what it had to teach, and not, when I came to die, discover that I had not lived. . . . I wanted to live deep . . . to drive life into a corner, and reduce it to its lowest terms, and, if it proved to be mean, why then to get the whole and genuine meanness of it, and publish its meanness to the world; or if it were sublime, to know it by experience, and be able to give a true account of it in my next excursion. For most men, it appears to me, are in a strange uncertainty about it, whether it is of the devil or of God, and have *somewhat hastily* concluded that it is the chief end of man here to "glorify God and enjoy Him forever." [1]

We wish surely to be among those who—face to face with the doubts and difficulties in life, with the shadows looming large over us—believe, nevertheless, that man can get some glimpse of the purpose and find in that purpose the way to the relief of doubt and to the lightening of difficulties.

[1] *Walden.*

Some people attempt to reach a solution by ignoring the evil and looking only at the beauty and the helpfulness. Such a solution need not concern us. There must be no slurring over of awkward facts. Could we entertain for a moment a theory of Divine Purpose which depended upon ignoring all the difficulties?

Equally unfortunate is the view that sees only the darkness and the pain, passing by all the sacrifice and altruism. We must take a true view of the facts—a synoptic view.

If we do this we must admit that the world is full of evil and that this seems to rule out a God of Love from the world, especially from the world of nature, where struggle and death seem so rife. We must not stop to consider the various ways in which men in the past have tried to orientate the existence of evil with the recognition of a God who is all-powerful and all-loving. It seems so simple a way out to argue that either God is not all-powerful, if He could not create the world without the evil, or that He is not all-loving if He could have done so and yet stayed His hand. The problem of life is not reducible to such a two-fold argument.

It has been maintained that biologists have been responsible for a pessimistic view of this problem, and, if we remember that the truthful recording of observations and of facts, however unpalatable, is one of the chief duties of the scientist, it is abundantly clear that many of the deeply cherished theories regarding both the inorganic and the organic world have been overthrown by scientists. This has caused them to be looked upon with suspicion. The dislike of awkward facts is usually visited upon the patient recorder of those facts, for it must be remembered that an old theory is more easily shattered than a new one formed.

No theories have suffered more severely from the iconoclastic shocks, caused by the flood of new facts set free by the group of Darwinian naturalists and their successors, than those relating to God and Nature, it is therefore illuminating and satisfactory to find that from modern biologists comes the main support of the philosophical arguments for the necessity of evil in any scheme out of which man, as a spiritual being, is to be formed. Let us then, before we plunge into the difficulties of the problem of evil, lay down briefly the ground upon which we stand; for it is my belief that from the standpoint of the evolutionary method of man's becoming, we shall be able to find a reasonable and satisfying answer to the problems before us.

We have endeavoured to focus attention upon the problem with man in the foreground. We have considered man and his place in Nature—a part of the whole. His solidarity and continuity are essential points to realise if we are to obtain any answer to our questionings. We can consider later the special problems associated with man's unique place in the scheme. We have tried to give an accurate picture of the evolutionary method, and have pointed out that in the drama of life, experimentation, trial and the overcoming which results by trying, are the best words in which to describe what has taken place and is taking place.

We see organisms moving about in a world governed by what we call fixed laws. That is to say, in the inorganic domain there is regularity, and this regularity is described by the recorders in shorthand labels for the various groups of phenomena. To state a law is merely to classify events, not to explain them. So when we say that the inorganic bodies are governed by law, we mean that given certain conditions the same result is always produced. Such a world of regularity is necessary for an evolutionary method, for evolution means that experience counts. There would be no value or meaning in experience if there was no regularity in the happenings of the physical world. Life is, as we have seen, a continual action and reaction, an interplay between the responsiveness of the organism and the pressures of the environment, a struggle with conditions which are fixed. An animal swimming in the sea has got the resistance and pressure of the water to strive against ; if there was no resistance there could be no swimming. A bird flying in the air depends upon the resistance of the air and the force of gravity ; if these conditions were not fixed, there could be no control. It is their very regularity that makes it possible for the conditions to be used. These are but two simple examples showing how organisms have learnt to use the environment, and serve but as a parable of the whole story. Experimentation, with increasing differentiation and integration, is the keynote, and the variability of organisms is the raw material for such experiments. The fact that in the presence of conditions newly appreciated, many of the experiments have come to nothing or have run riot, merely confirms the reality of the method. So too, the fact that some organisms have taken the easy line of parasitism, in no way invalidates the general argument of a developing system.

The age of reptiles, a period of great experimentation, brought forth the successful experiments of birds and mammals ; but it also resulted in many unsuccessful lines of development as the records in the rocks clearly demonstrate. It was worth it, however, for without this method the mammals would not have been evolved.

Another point of view must be kept carefully to the fore, this is " the web of life." No organism lives unto itself. Its struggles or slacknesses do not affect only itself, but groups of other organisms, altering their environment in one way or another. Animals do not live isolated lives, even if they are solitary in habit. We must not think of them as animals in a menagerie, caged off one from another, but living amid the storm and stress, the dangers and the difficulties of a real drama. The threads of their existence are interwoven with the threads of many others, so that no change in the life of the one can take place without disturbing—often very seriously—the lives of the others. Every new response made by an organism to the limitations and difficulties of its environment alters that environment, not only for itself but for others also. The threads that bind organisms together are many, and range the whole gamut from hunger to love, and we have seen that in the lives of many there are special linkages, which bring into play the new factor of sociability, with its wonderful results both individual and collective.

In this story of adaptation two kinds of hereditary registration are noticed. There is what is called natural heredity, the passing on of variations by means of the germ-plasm; and there is nurtural heritage, by which the experience of the family or herd is impressed upon the young forms. This last method is seen to be of great value in the higher groups when behaviour is more intelligent, and of greatest value in the group life of man, when the nurtural heritage becomes a code incorporating the successful experiences of social life.

Here we can see the gradual change of response and control from the instinctive to the intelligent, from the outward to the inward; for although the instinctive behaviour is still seen in relation to the fundamental needs and reactions of the organism, the intelligence and finally the intellect—intelligence become self-conscious—is able to initiate experiments, and thus to carry the animal-man more rapidly along his pathway to the spiritual man. One of the most interesting changes which the rise of intelligent behaviour has brought about, has been the enormous increase in experimental capacity, whereby the responses to the limitations and difficulties of the environment can take so many new paths; and when the brain power has reached the high development found in man, we see how many instincts are capable of being transmuted into other channels and to finer purpose. This change is not easy and the conflict is very real. The fact of being aware of the need for this inward control brings its difficulties, which often show themselves in a very real struggle for mastery. The very awareness seems to lead to rebellion as can be seen in the recapitulatory development of a child. "Why can't I eat some more?" "Why can't I do this?" "I will do it! I must do it!"

But the chief note that is sounded as we endeavour to catch the melody, which we are sure is being played, is the note of freedom or free-will. The only way to get a hint of the melody is to try to get a concerted and not a partial aural impression; and the treatment of the subject as outlined in the preceding chapters has been towards a wide and all-round appreciation of the subject. Viewed in this way we see that the fact that man is free to choose his way is the one great result of the age-long struggle of the evolutionary method. We must attempt to get this idea very clearly in mind for it is most important for a real understanding of the difficulties. We need not stop to argue whether man is really free, the question has been argued over and over again by philosophers in every age; we know, by our own experience, that he *is* free, and is not the mere creature of circumstance. We know that man has the power of choice. It may be that earlier choices have limited, in some measure, the immediate conditions that are open before him, but these conditions are the outcome of previous experiments and actions. It is true also that our inheritance and our environment have been limited by the choices in the past. This limitation, however, does not destroy the freedom, but is a real evidence that there has been the element of choice and therefore of freedom farther back. We know that the whole story of evolution has been possible by the refusal of organisms to be held down by the environment. The conquering note makes no uncertain sound, and the failures—or what we call failures, when we see from man's point of view the process as a whole—do but prove the thesis of freedom. If everything was determined, if organisms were machines, the whole drama would be meaningless, and we should be faced with a bigger problem than we have at present. If a mechanical universe was the purpose, with men as highly organised types of machine, then any solution of this evolutionary struggle we have been investigating is impossible, except along the lines of diabolic frivolity. If machines are desired, there is surely no need to use the wasteful method of evolution. We need not continue these suppositions, for, as we have seen, the facts of evolution are clear, and the facts of evolution, in spite of difficulties and unexplained details, point clearly to the winning of freedom. Man is not a machine, nor is amœba a machine, nor is a developing egg-cell a machine. The whole story is a-throb with purpose and the purpose is the rise into self-determination.

Freedom cannot be given. To think of freedom

as a gift involves a contradiction. Freedom must be won, and the winning must start far back at the very beginning; it must start in the action and reaction between organisms and their inorganic law-bound environment. These conditions of regularity in the inorganic domain are the necessary ground-work for the life of organisms, and as such are an essential part of the scheme.

But freedom becomes ever a finer freedom, until when man, a spiritual being in touch with the Spiritual Source of Power and Life, is able to make a conscious choice, and stand for the good, the beautiful and the true in face of all the consequences, coming into the freedom for which he has been destined, the freedom of a Son of God.

The choice is not a conscious one in the lower forms of animal life; but we have already seen that, even in the simplest animals, there is something inherent that prevents equilibrium, something that is creative, that is not bound by mechanical laws, and that refuses (without knowing that it does so or why) to be held in by its immediate environmental pressures, but is continually reaching out towards a fuller and a freer environment. This inherent, controlling power becomes more and more evident as the animal becomes more elaborate and more capable of reacting to fresh parts of the larger environment. Eventually it shows itself in the form which we recognise as a mixture of instinct and intelligence, and from this general consciousness can be traced the self-consciousness of man. Here the direction of choice and the value of choice can be realised, and the animal becomes a moral agent. He thus becomes conscious of new influences, other features of his environment to which he is able to respond, and by this response, know and control and win through to finer ends.

The point we would make here is this—taking for granted that freedom is the outcome, the kind of world in which we find ourselves is the only possible kind. Freedom could not have been obtained otherwise. The world is an evolutionary world and it has brought forth man with power to choose. Any other method of creating man would have failed in its purpose, for the scheme must be such that real freedom is a possibility in the end and this is only possible if freedom could be won from the beginning. Our world is God's handiwork, and a real expression of His nature. The end of it all conditions the method of its creation, for in God there can be no contradiction. Motive and method have to be in harmony. If freedom is essential, then the method must be such as to bring forth freedom. Life is a great adventure, and the conditions of our world, of our environment (inorganic, organic, and spiritual) are such that the adventure is one long climb towards goodness. Goodness is not possible without freedom.

We have considered the ground of our becoming and we must now carry our enquiry beyond the bounds of biology to ask why freedom is so essential? Man is free to choose. Is there any indication of the meaning of freedom to bring forward to help us in our study of the problem of life? This is a question for the philosopher, and in indicating some lines of thought I do not wish to lay more claim to this title than can be claimed by any ordinary man who is trying to obtain a reasonable answer to his queries about life. I approach these problems as a biologist, and I have attempted to give an unbiassed account of the matter; but one cannot stop short at the end of the biological argument; thoughts and ideas run forward to join hands with thoughts and ideas reaching downward from the side of religion. The evolutionary thinker merges into the religious thinker. It remains to be seen whether the result is worthy of consideration; but the discussion is continued with much diffidence and with as little dogmatism as possible. In face of the burden of evil which overwhelms so many people as an insoluble problem, or as a blot on the scheme, we unite in those words of Josiah Royce, condemning superficiality and half-way answers given to those whose lives are under the stress and strain of sorrow or of sin: "They deserve either our simple silence, or if we are ready to speak, the speech of people who ourselves inquire as Job inquired."

* * * * *

Man is free to choose because only so could man become a moral being, and a moral being is one who, deliberately, of his own free will denies the selfishness which seems to press upon him. Goodness is a spiritual activity and, like all other kinds of activity, has no meaning unless the action is against something constraining and retaining. The activity of a bird is not possible without the resistance of the air and the pull of gravity; in the same way this higher kind of activity which leads to character is not possible, except in an environment of motives and dispositions which have to be overcome and thereby used in the rise out of the lower into the higher life. Man is conscious of this conflict. He sees that he is called to stand up against his difficulties. Only so can he come into a further freedom, only so can he begin to tread the pathway to goodness, and to have a personal realisation of the satisfaction which comes as his will enters into harmony with other individuals and with the Spirit of God. By exercising our spiritual activity, or by striving after goodness, the most important part of this activity, we find greater possibilities opening up before us. We begin to realise the end towards which the scheme is moving, as those "who, rowing hard against the stream, see distant gates of Eden gleam and do not dream it is a dream."

There is somewhat within us which finds satisfaction in doing our duty, not because we like it, but because we feel the call to nobleness for its own sake. When faced with a difficult question, I can, if I like, tell a lie, that is a sign of my freedom; but in refusing to do so, I am exercising my spiritual nature in its growth in morality, a part of goodness. Judged as an immediate policy, honesty is not the best policy, as policy pure and simple, a lie would save a lot of trouble; but the act of truth-telling, although it will need a struggle, produces a feeling of inward instinctive satisfaction or harmony. It matters not that the psychologist calls the feeling a psychical accompaniment of instinctive behaviour, for we have already seen that awareness of union with God is largely a matter of the intuitive or inward part of our consciousness. The evolutionary world has been the training ground for man, and man has been able in such a world to achieve freedom and so to achieve goodness, and to come into conscious harmony with the purpose of the scheme.

But does not the scheme leave open the possibility of the triumph of evil just as much as the triumph of goodness? In allowing man to make himself, is there any guarantee that the result will work out as we have indicated? We shall return to this problem in a later chapter; but there are one or two considerations which seem to me to indicate that good can triumph, if man is true to his nature. "We are so truly made, if only to our making we were true," as John Drinkwater puts it. Taking a long and broad view of the facts, history does show a rise in moral values. There has been progress. Evils have been and are being "overcome." Goodness in man is such that it makes goodness easier and more certain for others—its power is cumulative; whilst selfishness is destructive and is a disintegrating force. Goodness in practice brings strength and a greater power in doing good; whilst evil-doing brings a slacking of effort and a weakening of character. Goodness awakens in us a sense of fitness, we respond to it. The inward intuitive consciousness is stirred by goodness, as it is stirred by beauty; whereas this same inward spiritual nature is repelled by bad actions. We must beware of arguing in a circle, but it seems as though the exercise of the moral activity brings us into harmony with an environment which we recognise as all-good, even though our recognition is dim and uncertain. Personal experience convinces us that our freedom finds its true meaning when it is exercised in accordance with the belief that the purpose of the world is the triumph of goodness; and that our deepest satisfaction comes as we enter into a loving partnership with the Purposer.

* * * * *

We turn now to some of the difficulties which are grouped under the general title, "the problem of evil." I do not wish to overlook any difficulty or to give superficial answers to the questions that arise, but, if the evolutionary concept has been at all adequately realised, it will be seen that a number of the shadows are inherent in the system. Take, for instance, the idea of "the web of life"; its nutritive and self-preservative threads travel in all directions, experimentation leads to the ant-hill as well as to the tape-worm. Specialisation pushed too far in one direction, results in extinction or failure to progress. The apparent wastefulness seen in the vast numbers of organisms, and the nutritive relationships which seem to mean so much cruelty, are in less danger of exaggeration, and are more understandable when the drama is seen as a whole. Most of these difficulties are due to partial views. Then there are many difficulties caused by man's stupidity, greed and selfishness. These must not be laid at Nature's door, even though man has often tried to excuse his actions by attributing to them the sanction of Nature's method. The greatest and most widespread of these man-made evils is war, and we cannot pass it over without some attempt to see clearly the causes that have produced such a blot upon civilisation, but first of all we must discuss the central problem.

The evil in the world is usually grouped under two heads, physical evil and moral evil, or more simply, suffering and sin. They are both very real, terribly real, but they are both understandable in the light of our discussions upon the method and purpose of evolution. Physical evil is inevitable in an evolutionary world. Moral evil is understandable, but although it is easy to see the causes of its presence, it is not inevitable. Physical evil should gradually become less, as man learns to control the forces of nature. The problem of moral evil becomes the personal conquest of sin or, to give it a positive form, the personal triumph of good. There is always the possibility of sin and this possibility is inherent in the evolutionary method. Sin is possible because man comes to realise good from evil, and, facing the alternatives, he has the power of conscious choice. It is not a toss-up. If it were, then in the long run the good and evil actions would balance one another. No, the choice has more in it than that, and man is not without a guide, both inward and outward, to point out the steep ascent.

A great deal of the pain and suffering in the world is simply due to the fact that, without danger, no one can be exposed to risks. The danger is in the scheme. The physical environment works along lines of regularity, and the life of the organisms is one long struggle in overcoming. But if there is to be overcoming, there must be something to overcome. If effort is the law of life, if experiment is necessary in order to obtain experience, then the possibility of disaster is a certain accompaniment. Freedom for goodness cannot be obtained in any other way; we must, therefore, face the burden which the method entails. Some people wish to take all the advantages of the evolutionary method and none of the disadvantages. A little thought will convince us that to expect the suspension of the operation of the primitive cosmical forces for our private convenience or safety is arrant selfishness, as well as asking God to contradict His method of creation. The method, as we have seen, depends upon the trustworthiness of the cosmical forces, and any suspension of the working of gravity to save man from his or from someone else's carelessness or ignorance, would frustrate the whole scheme and bring chaos and confusion. The force of gravity is there to be controlled, as it is controlled and overcome every time an arm is raised. The fact that a biıd is able to defy the pull of gravity depends upon the use of its own powers, this fact will not prevent it from falling if those powers are not exerted. To abrogate the laws of diffusion and density, when the carelessness or ignorance of man brings about a *Titanic* disaster, in order to save men from sinking and drowning, would be a contradiction of Divine method. Water has certain properties, it has a specific gravity due to its physical composition and it has the power of holding oxygen in solution, these properties are taken advantage of by the fishes and lower aquatic forms of life; but when animals climb out of the water into the environment of air, the new conditions call forth new adaptations, and these very adaptations, as they become perfected in the evolutionary way, make a fall into the water a serious danger.

Many physical evils have disappeared before the controlling hand of man, yet we stand aghast at the vast amount of preventible suffering there is all around us. Failure in our knowledge, failure in our application of new knowledge, slackness, and distrust of new ideas; all these things add to the pain and suffering in the world. This is not inevitable, it is preventible, and its presence indicates one of the many phases of evil. There are, then, the natural disadvantages, which are necessarily present as a result of the purpose, for without them there would have been no evolution and no progress; but the amount of pain and suffering seems to many to be much greater than is needed to form a training-ground for the rise of freedom. Who can say how much is necessary? Who is ready to put a limit and to say "thus far and no farther"? We know too little of our world with its purpose of freedom, to offer anything approaching a final opinion upon the matter. It may be that the method is such that any control would have vitiated its purpose, and that the interplay between organisms and their environment must not be restricted in any way. At the same time we must not over-emphasise the problem. The world is not " a hymn of love," nor is it a gladiatorial display, and we do a disservice to Nature by projecting our man-formed ideas of pain and suffering upon her.

It is uncertain to what extent, if any, pain is a part of the problem among the lower forms of animal life. In creatures whose behaviour is instinctive, it is difficult to see how pain can play any part in their consciousness. Even in the lower mammals, with a much higher type of nervous system, and with their behaviour governed partly by intelligence, there is little pain, but much fear. Sensitiveness to pain is associated with the most highly organised animals, and reaches its highest development in man; but we know, too, that men differ remarkably in their sensitiveness to pain, and any one individual will vary in his sensitiveness according to the circumstances. The power of suggestion and its effect in causing and relieving pain is one of the most certain facts of modern medical knowledge. In the interplay of an animal of well-developed intelligence with its environment, the consciousness of pain is necessary in order to avoid injury. The higher the animal—that is to say the more fully an animal is responding to the environment

—the more important does pain become as a guide and protector from injury. An animal whose behaviour is entirely instinctive, whose actions are automatic and have not to be learnt by the trial and error method in the animal's own experience, would have no use for pain; and it is probably that those animals—such as the insects—which specialise in instinctive behaviour do not feel pain. Death is usually the immediate result of any failure in behaviour. In man the highly organised body and nervous system enable him to steer his life amid the manifold dangers of his path, and the sensitiveness to pain is the method. Thus pain is part of the scheme, and a necessary part, and is closely associated with the rise of intelligence. The rise of intelligence is a step towards self-consciousness and self-determination. In man the sensibility to pain becomes the mental problem recognised by the term "suffering." The chief sources of pain are not physical but mental troubles, such as fear, anxiety, disappointment and remorse.

Another great source of pain is the sin of man. No man can sin unto himself. The web of life is very real, and the threads that bind us together are many and varied. A vast amount of suffering is caused to others by our failures; but there is the other aspect of the web. Although by the ties of our lives we share in the results of failure, yet by virtue of the same ties we share in the results of victory. The power of man in rising above the suffering and in transmuting it by his creative spirit is not isolated in its action, but spreads in helpful circles, encouraging and raising others. We must return in a later chapter to consider this power to victory in its relation to the Providence of God.

We have now to consider how the other form of evil, moral evil or sin, is an outcome of the scheme. It will be very evident that, taking the standpoint which has been developed in the preceding pages, the problem of sin is much more easily understood than the broader problem of physical evil. Sin is the conscious failure to come up to the standard which we know to be right. It is a failure to continue the evolutionary struggle. It is the lower choice when a higher one is before us. It is the failure to become what we might have become. Sin is, of course, the bye-product of the scheme. The scheme did not need men to sin; it made sinning possible, because the scheme has as its purpose the winning of freedom. Freedom to love and serve, the acquisition of self-mastery for service and the growth into a moral agent which is involved, need an evolutionary world in which, with effort as the law of life, the following of the right is possible only because the following of the wrong is also possible.

The experiment of following the right began far back in the early days of man's social life, when dead against his inclinations and instincts, against custom and prudence, some man did a kind or good action (that is kind or good as judged from his standpoint). This must have been a great step forward, and a new kind of satisfaction must have come into that man's consciousness;

not the satisfaction which he had been used to, but a deeper, a more inward and personal satisfaction. Is all of this pure supposition? I think not, for observations upon the development of the moral sense in children and savages furnish clues, and in our own experience we are conscious of this same satisfaction, however dimly it may be appreciated at first. Ideas of right and wrong have developed rapidly in relation to the advancing intelligence of man, for the intellect finds ever fresh fields for the combat.

Perhaps the easiest kind of sin to recognise has to do with our instincts. Probably that early ancestor of ours, feeling hungry, restrained his own desire in order to let another eat, another one more in need of food than he, or less able to get it. Our passions and instincts may become sources of sin if we do not control them, if we exhibit them as animal instincts instead of instincts of animal origin man-controlled. Hunger, for instance, is not in itself sinful, it becomes sinful if it is allowed to lead us into actions which we know are not right.

Some sins are due to a failure to recognise the wider group. What is recognised as wrong within one's immediate social group is looked upon as right or allowable if a member of another group is the victim. To take our case of the hunger instinct. To steal food from a fellow-member of a clan would be a sin, but not if the stealing is from a member of a rival clan. Sins may be anti-social, but many are not condemned because our group is too narrow a one. Out of this failure to recognise the wider group rise many of the modern social and religious problems. A man who would not dream of stealing a penny from his next-door neighbour may be found defrauding, without a qualm, the public revenue or the railway company. This limitation of the group, within which our morality or religious feelings operate, is one of the great factors in the delay of the purpose, and a very potent cause of sin and suffering.

Sin is man-made; it has been allowed by God because unless man could choose the wrong, the value of choosing the right would be nil. To be incapable of sinning, man must be a sensitive automaton, an animated puppet, not a man. Sin, then, is a sign of man's freedom, is evidence of his creative power, and shows that his power of choice is a real choice. From this point of view the possibility of sin is an argument for the goodness of God. It is a measure of God's trust in man—a trust woefully broken and denied—and yet because of this very trust man is able to appreciate what his freedom means, he is able to see the Love which risks all for the creation of love, and, seeing the glory of this vision, he is ready to strive to do his part in the scheme.

Thus both aspects of evil are bound up with the scheme, and, although we have not faced all the difficulties and have only taken a general view of the problem, yet I do believe that, from the biological angle, the shadows are seen, no longer in isolation but as part of the whole, and the whole is beautiful and good.

A few lines from an essay by Dr. F. R. Tennant are worth quoting, for they deal with the two chief objections to the line of argument outlined above and serve to summarise the main thought.

> The objection to Theism on the score of physical evil is that there is too much law or regularity in the world; the objection on the score of moral evil is that there is too much contingency in the world. And these objections are plainly contrary, the one to the other. Safeguards which would make us immune from temptation might secure objective right-doing, such as would possess absolutely no moral worth; they could not secure the morality which in the last resort is love—the only real fulfilling of the law—and they would not conduce to moral strength or moral character. Character is made, not born; and it must be self-made, not received ready-made. The development of morality is naturally not continuous or uninterruptedly progressive; but to coerce it to be so would be to destroy its morality altogether.[1]

Browning also expresses a similar view in *The Ring and the Book* in the following lines:

> I can believe, this dread machinery
> Of sin and sorrow, would confound me else,
> Devised—all pain, at most expenditure
> Of pain by Who devised pain—to evolve
> By new machinery in counterpart,
> The moral qualities of Man—how else ?—
> To make him love in turn, and be beloved,
> Creative and self-sacrificing too,
> And thus eventually Godlike.

In this chapter we have done little more than attempt to clear away the doubts about the good purpose in the scheme; the problem of man's

[1] *Pain and Conflict in Human Life*, Camb. Lectures, 1916.

part in the triumph of good is to be considered later. We have been standing outside as spectators of the drama of life, attempting to understand man's place in it. The fact that we are able to do so marks the uniqueness of man, but I do not agree when E. Ray Lankester calls him " Nature's rebel." He should, rather, be called Nature's conqueror and king. The note of struggle and rebellion has been dominant all through the story, and rather less than more in man's case. The kingdom of man is the work of man, and we shall have to try to understand how this is to become the Kingdom of God.

I had thought to end this chapter here, but I feel constrained to deal with one further difficulty. This shadow is not a part of the scheme as such, although, being man-made, it might claim a place in our consideration. I refer to war. It is because the war has brought many people face to face with the problems of evil in an intimate and personal way, that I have decided to discuss the subject here rather than in a later chapter. The problem of war in relation to human life has been variously stated. Is war a biological necessity ? Is it a real feature of the lower environment out of which man is climbing, and a necessary, though an abhorent, phase in his development towards goodness ? Can we look upon war as a part of what is called "the struggle for existence" ?

What is war ? We think naturally of the Great War. It is a struggle between fellows of the same species, and a struggle in which there are no rules. Death, injury, destruction as the means to victory are the objects of both sides, and to accomplish the end any device is allowed. The engines of warfare are not the claws and teeth of the combatants, but the wonderful mechanical creations of man's mind, whereby death and destruction can be accomplished at great distances. With the assistance of chemistry and bacteriology the destruction can be still more extensive. The distinction between combatants and non-combatants has broken down with the long-range guns and aeroplane warfare, and is no longer possible. Women and children suffer both directly and indirectly.

In this generation there is no need to give details of war's reality. Its filth and horror have burnt deep into the hearts of many, and, although the reality is too dreadful to parade in any fullness in verse or prose, yet those who have not experienced it can get something of the truth from *Under Fire*, by Henri Barbusse, and the many similar books now published.[1] War shows no discrimination in those killed and those allowed to escape. It seemed as though the very bravest and best of our young manhood were the ones to be killed off. There is no desire in this to compare personalities or to suggest that the weaker ones of the race have not a place in the common life to which we are called. The point to make clear is that considering war from a biological point the selective value is negative and

[1] *The Grey Wave*, by Major Gibbs, can be warmly recommended.

that it works in the opposite direction from the selective process in Nature.

Taking man as an animal, can we say that war has been a factor in his progress or likely to be a factor in his further development ? The fact that struggle has been the keynote to advance and that effort is the law of progress, lead some to assume that because war is a struggle, it must therefore be part of the scheme.

We have discussed, in an earlier chapter, the very limited ideas which many have about the struggle for existence, and this limitation has had a very serious result in furnishing a plausible argument for war. Even among animals low down in the scale of animal life, where it is assumed that mutual destruction is the normal state of things, there are no facts to support this general assumption. Struggle between fellows is a very small part of the "struggle." The real meaning of the term "struggle" lies rather in its use for all the ways in which a creature responds to the limitations and difficulties which surround it. Among the higher animals, those in which intelligence enables a more varied response to be given to the limitations, we find less and less signs of mutual struggle and more and more signs of mutual aid and sociality. So that we can say, regarding man simply as a member of the Primate mammals, considering his structure and his habits, war has not been a factor in his evolution.

Of course a calm examination of war itself would bring anyone to that conclusion, for to be of any value in a scheme of animal evolution which depends upon variability and the selective sifting of Nature, war would have to show that it eliminated the anti-progressive and the anti-social members, and directly favoured the qualities which we have called good. War is indiscriminate in its action, and even in the earlier days when hand-to-hand conflicts were the order of the day, there was no guarantee that physical strength always combined with the higher psychic qualities. It is conceivable that animal courage and brute force might have been selected, but the life of the social animal does not depend so much upon animal strength, as upon the peaceful qualities of sympathy and altruism, although there is ample scope, in putting these into use, for the exercise of a courage none the less virile. In modern war even the possible survival value of brute strength is lost, and the appeal of war to biology for support must fail.

A consideration of history, too, especially if we go back to the very early records of man's life on the earth, fails to support the evolutionary value of war. Indeed, war is seen to be a comparatively late acquisition. It has arisen as a bye-product out of his social arrangements, due to a change in habits, and war is in no way a cause of his evolution.

Early man was a hunter, not a fighter, and we can see how gradually the smaller groupings into families and clans led to jealousy, desire for revenge, plundering and war. Probably the more permanent settlements and the growth in the idea of property and of hunting rights had a good deal to do with the rise of war.

As Trotter puts it,

> Man's complete conquest of the grosser enemies of his race has allowed him leisure for turning his restless pugnacity—a quality no longer fully occupied upon his non-human environment—against his own species. When the major units of humanity were small the results of such conflict were not perhaps very serious to the race as a whole, except in prolonging the twilight stages of civilisation.[1]

I am spending a little time upon the biological side of the problem, because it is necessary to clear this argument for war right away, so that the other arguments based upon man's unique position in the scheme and upon his mental and spiritual nature can have their full weight. Whilst there is any doubt in one's mind about the value of war from an evolutionary point of view, the arguments from the other side can be brushed aside or ignored.

> Perhaps no general expressions have been more common since the beginning of the war, in the mouths of those who have undertaken our instruction in the meaning of events, than the propositions that pacifism is now finally exploded and shown always to have been nonsense, that war is and always will be an inevitable necessity in human affairs as man is what is called a fighting animal, and that not only is the abolition of war an impossibility, but should the abolition of it unhappily prove to be possible after all and be accomplished, the result could only be degeneration and disaster.

[1] *Instincts of the Herd*, p. 130.

Biological considerations would seem to suggest that these generalisations contain a large element of inexactitude. The doctrine of pacifism is a perfectly natural development, and ultimately inevitable in an animal having an unlimited appetite for experience and an indestructible inheritance of social instinct. Like all moral discoveries made in the haphazard, one-sided way which the lack of co-ordination in human society forces upon its moral pioneers, it has necessarily an appearance of crankiness, of sentimentality, of an inaptitude for the grasp of reality. This is normal and does not in the least affect the value of the truth it contains. Legal and religious torture were doubtless first attacked by cranks; slavery was abolished by them. Advocacy by such types does not therefore constitute an argument of any weight against their doctrines, which can adequately be judged only by some purely objective standard. Judged by such a standard, pacifism, as we have seen, appears to be a natural development, and is directed towards a goal which unless man's nature undergoes a radical change will probably be attained.[1]

Dr. P. Chalmers Mitchell, the Secretary of the Zoological Society, is one of several biologists, who have examined in detail the claim of war to be considered biologically necessary. His *Evolution and the War*, is a valuable contribution. A short extract must suffice.

Looking through the Animal Kingdom as a whole, and remembering that the Vegetable Kingdom is as much subject and responsive to whatsoever may be the law of organic evolution, I find no grounds for interpreting Darwin's 'metaphorical phrase,' 'the struggle for existence,' in any sense that would make it a justification of war between nations. It is my business just now to refute a misconception of the struggle rather than to explain

[1] *Instincts of the Herd*, p. 125.

what it is. But, if the latter were my task, I could adduce from the writings of Darwin himself, and from those of later naturalists, a thousand instances taken from the Animal Kingdom in which success has come about by means analogous with the cultivation of all the peaceful arts, the raising of the intelligence, and the heightening of the emotions of love and pity.

War, then, is a gigantic sin. It is a man-made evil and it is not a necessity. It is no argument in its favour to show that it has given a chance for the exercise of fine qualities, and that love and self-sacrifice have been prevalent. If these things have any argumental value, they will serve as an indictment of our social life and of our religion. The normal life has been too much robbed of its spirit of adventure, and the spirit of slavery (the economic variety) has held men in its degrading grip.

There are many other arguments against war, and some of these have been put forward in my book "*As a Man Thinketh* . . . ," and I do not propose to retravel that ground again here.

War has held man in its grip too long to be easily thrown off, and the call comes to all who believe in the sinful character of war to help on all the various plans which are being put forward for its prevention.

These plans may concern the external arrangements whereby the social groups can be widened and the threads of friendly inter-relationship increased ; or they may concern the inward life of man. If only we can realise that war is the result of sin and that sin is a wrong response, and that the wrong response is an inward action before it expresses itself in outward form, we may see the way to its conquest. War cannot destroy war. War must be destroyed in the heart. This is a personal matter and the call to live the higher and truer life of love is not only a call to struggle against war; it will also call us out against the other man-made and man-allowed evils which disfigure our civilisation.

* * * * *

We have been facing big difficulties. Can there be a purpose in face of all this evil? Can that purpose be good? We have attempted to travel the way towards an affirmative answer. The way, I believe, is there, and each in his own life can realise the truth of this conception of the purpose.

The thought with which I would end this chapter, is the wonderful thought that the author of the scheme is also Himself in the scheme, and that He has been suffering and is suffering by reason of the evil in the world. If this is true, and I believe it is, then what an added value is given to goodness and how strong should be the call to faithfulness.

I take in conclusion, a few paragraphs from an essay on Job written by Josiah Royce:

> If moral evil were simply destroyed and wiped away from the external world, the knowledge of moral goodness would also be destroyed. For the love of moral good is the thwarting of lower loves for the sake of the higher organisation. What is needed, then, for the definition of the divine knowledge of a world that in its wholeness is perfect, is not a divine knowledge that shall ignore,

wipe out and utterly make naught the existence of any ill, whether physical or moral, but a divine knowledge to which shall be present that love of the world as a whole which is fulfilled in the endurance of physical ill, in the subordination of moral ill, in the thwarting of impulses which survive even when subordinated, in the acceptance of repugnances which are still eternal, in the triumph over an enemy that endures even through its eternal defeat, and in the discovery that the endless tension of the finite world is included in the contemplative consciousness of the repose and harmony of eternity.

* * * * *

The existence of evil, then, is not only consistent with the perfection of the Universe, but is necessary for the very existence of that perfection.

* * * * *

Your suffering, just as it is in you, is God's suffering. No chasm divides you from God. He is not remote from you even in his eternity. He is here. His eternity means merely the completeness of his experience. But that completeness is inclusive. Your sorrow is one of the included facts.

* * * * *

But to the serious moral agent we say: What you mean when you say that evil in this temporal world ought not to exist, and ought to be suppressed, is simply what God means by seeing that evil ought to be and is endlessly thwarted, endured, but subordinated. In the natural world you are the minister of God's triumph. Your deed is His. You can never clean the world of evil; but you can subordinate evil. The justification of the presence in the world of the morally evil becomes apparent to us mortals only in so far as this evil is overcome and condemned. It exists only that it may be cast down. Courage, then, for God works in you. In the order of time you embody in outer acts what is for him the truth of his eternity.[1]

[1] *Studies of Good and Evil*, Josiah Royce, p. 24 *et seq.*

BOOKS

Halliday, W. Fearon	Reconciliation and Reality.
Micklem, N. . . .	The Open Light.
Mitchell, P. Chalmers	Evolution and the War.
Nicolai, G. F. . . .	The Biology of War.
Streeter, Canon B. M. (Editor)	God and the Struggle for Existence.
Thomson, J. Arthur .	Biology and War. Papers for War Time, No. 24.
	Darwinism and Human Life.
Trotter, W. . . .	Instincts of the Herd in Peace and War.
Unwin, E. E. . . .	"As a Man Thinketh. . . ." The Personal Problem of Militarism.

CHAPTER VI

THE APPEAL OF BEAUTY

" Every blade of grass, each leaf, each separate floret and petal, is an inspiration speaking of hope. . . . My heart is fixed firm and stable in the belief that ultimately the sunshine and the summer, the flowers and the azure sky, shall become, as it were, interwoven into man's existence. He shall take from all their beauty and enjoy their glory. Hence it is that a flower is to me so much more than stalk and petals."

RICHARD JEFFERIES.

" It was of high value to look upon beauty, not as a stranger in the world, not as a causal aspect afforded by some phenomena under accidental conditions, but as the fortunate revelation of that principle which permeates all reality with its living activity."

LOTZE.

ANY consideration of the philosophy of biology and of the relationships between the world of Nature and man, its crown, would be incomplete without touching upon the fact and the meaning of beauty. It is not an easy subject to discuss, though the difficulties are very different from those we have been considering in the preceding chapter. In the problem of evil, disharmony is the prevailing note ; in the problem of beauty it is the harmony that gives it the appeal and the difficulties are in the very inadequacy of language to give any satisfying account of it.

The fact of beauty in Nature is recognised by everyone in a greater or lesser degree. We have all some appreciation of the form of clouds and mountain masses, or of the group of trees seen against the sunset ; we are all stirred in some measure by the colours of sea and sky, and of the flowers and leaves ; we all follow with admiration the skimming of the swallow and the easy grace of the sea-gull. The sense of the beautiful, too, can be developed, and we are all conscious of growth in discernment. At first we may be prone to look for the spectacular, the Alps, the rainbow, the stretch of purple heather, the group of red deer ; but as our appreciation widens and deepens, we find stirrings in our spirits from the common things of life, and we learn Peter's lesson, that nothing is common or unclean. The pebble by the wayside, the carpet of dead leaves, the feather dropped by the moulting bird, the dew-spangled leaf, each in its turn may call us with its note of beauty and become to us a sacrament.

> The tree (says Blake) which moves some to tears of joy, is in the eyes of others only a green thing that stands in the way.

The more our knowledge of animal and plant life grows, the more scope do we find for the exercise of our appreciation of the beautiful ; the more we probe into the intricacies of structure the more beauty is revealed to us. No one who has ever seen the microscopic structure of a blade of grass or a sea-urchin's spine, could fail to be impressed by the pervasiveness of beauty in Nature.

If we begin to ask ourselves of what the beauty consists, we come at once up against the difficulties of the subject, and shall find ourselves unable to give any adequate answer. In regard to many cases something can be said, for we can see that the appeal comes from form or colour or movement, or a combination of form and colour. In other cases we appreciate a sense of fitness in the usefulness of form or colour to its purpose in relation to the life of the organism, especially in some cases of artifice, or we see a certain rhythm of growth. But in many cases there is nothing to say, save that there seems to be in the beauty something of an innateness, something implicit in the very substance of the organism, something that is there without use or any understandable purpose. In general, we might say that beauty is exhibited when a sense of harmony is awakened.

The philosopher defines the beautiful as that which excites in us an emotion which is known as the Æsthetic Emotion. This emotion is not like any other feeling; it is different from the sensual feelings, although it may awaken thrills of joy which can be detected in their bodily reactions. In a large majority of cases it is caused by the perception of objects, which objects raise within us this special feeling. Without the objects or the memory of the objects the emotion would not be stirred.

What is Beauty? Many have asked it, and could find no answer, because they understood their question no more than jesting Pilate understood his 'What is Truth?' But many beside have asked it with at least a real desire to understand. It was already in the mind of the prehistoric artist who was the first to draw a pattern or to sketch the mammoth, though no doubt he did not put the question to himself. It has been there, expressed or unexpressed, wherever a man has had vision enough to find his spirit stirred by a flower or a cathedral; a fabric or the low October sun upon a sheet of gossamer; wherever a man has tried to reproduce nature on canvas or pour out his longing and triumph in sound or written words. He has cried out that beauty dwells only in his own spirit, for there have been moods and days when he could see no beauty in that which at other times moved him deeply. Yet the agreement of civilised mankind, at all events, that this or that particular is beautiful is so widely diffused, that he cannot but admit that something in the object itself must suggest the idea of beauty. Taste may change, but the sunset and the rose are universally acclaimed by all who have any æsthetic perception at all.[1]

Some people argue that the idea of beauty is wholly in our minds, that it is subjective. That the state of our mind affects our appreciation we know only too well, but if we attempt to separate our ideas of beauty from the objects which have caused them, we shall find but little idea of beauty left. The object is needed for the idea to arise. The emotion is subjective, but it did not arise out of nothing. It was something in the object, which was realised by the instinctive part of our consciousness. We need not go into

[1] *Beauty and the Beast*, S. A. McDowall.

the arguments put forward by the exponents of the subjectiveness of beauty, for the all-pervading beauty in Nature, the universality of its appeal and the fact that man has been cradled in Nature —is, indeed, the child of its very soil—seem to supply the answer to our queries. It is not surprising to the biologist that the revelation of beauty in form, in colour, in adaptativeness, in movement, have left their mark on the spirit of man. The response to light in the development of eyes is a parable of the æsthetic disturbance, set up by the perception of beauty; it is like another gateway into our personality.

Let us turn for a moment to the evidence of men who have felt the call of the beauty in Nature, and have used it, in Lowell's words about Thoreau, "as a mountain-path to an ideal world." These poet-naturalists, as they are sometimes called, give us some hint of the uplift that comes into their lives by the stimulus of Nature's beauty. They hold communion with her and she speaks into their inmost hearts. They have the same difficulties as the other mystics, they tread their pathway to Reality—the vision of the ideal—but their footsteps cannot be adequately mapped. They have their intuitions, their spirits are touched to finer issues by the beauty around them, they feel that they are in the presence of Reality; but their expression of this emotion in the vehicle of language is inadequate to their feelings. Besides, their outward expression is conditioned by the other forces which play upon the mind. Education, upbringing, the circumstances of their lives, their own peculiarities, all these things and many others, will have an effect upon the form of their expression. A Gilbert White will be different from a Meredith, a Ruskin will be different from a Richard Jefferies. There is no doubt in their minds of the reality of their emotions. They are not in the throes of pure imagination. They are conscious also that they have been influenced by outward objects, and they have no doubt about the beauty of those objects. Wordsworth is in no doubt that the pleasure which fills his heart had been caused by real daffodils, growing alongside the lake under the trees, and all of us who have seen them growing in their natural habitat agree with him, and, like him, can thus come into touch with spiritual reality. That is to say, the beauty of the daffodil is a code through which something of a spiritual meaning comes home to us. Ruskin, in a quaint passage, emphasises from the artist's point of view the value of form. Speaking of scientific philosophers, he says:

> They come and tell you, for instance, that there is as much heat, or motion, or calorific energy (or whatever else they like to call it) in a tea-kettle as in a Gier-eagle. Very good; that is so; and it is very interesting. It requires just as much heat as will boil the kettle, to take the Gier-eagle up to its nest; and as much more to bring him down again on a hare or a partridge. But we painters, acknowledging the equality and similarity of the kettle and the bird in all scientific respects, attach, for our part, our principal interest in the difference in their forms. For us, the primarily cognisable facts, in the two things, are, that the kettle has a spout and the eagle a beak; the one a lid on its back, the other a pair of wings.

Ruskin, of course, goes on to point out that in these wings we find a beauty, and the beauty awakens sensations within us which would not arise except in response to the form. "Hold fast to the form and defend that first, as distinguished from the mere transition of forces. . . . It is curious how far mere form will carry you ahead of the philosophers." Ruskin saw in beauty the sacrament of the Beauty of God, and thus the value of beauty in relation to human conduct was his great message.

Richard Jefferies is all the time painting wonderful word-pictures of the objects around him, because these objects meant so much to him. He felt that the beauty of these objects had a message for his inward being, and he desired to communicate this same message to others, by bringing them by his words into a similar relation with the beautiful objects. Just as the words of his essays convey in some measure his meaning, so the beauty of Nature is code revealing something of the spiritual mysteries to man.

> Never was such a worshipper of earth. The commonest pebble, dusty and marked with the stain of the ground, seems to me so wonderful; my mind works round it till it becomes the sun and centre of a system of thought and feeling.[1]

And again,

> Through every grass blade in the thousand, thousand grasses; through the million leaves, veined and edge-cut, on bush and tree; through the song-notes and the marked

[1] From "Hours of Spring" in *Field and Hedgerow*.

feathers of the birds; through the insects' hum and the colour of the butterflies; through the soft warm air, the flecks of clouds dissolving—I used them all for prayer.[1]

What do we experience when we are touched by the beauty of Nature? Let us again ask these worshippers of the beautiful to explain their rapture. We shall find their message expressed in very differing language, but the message itself has a unity about it that speaks of reality. They feel around them the presence of an Unseen Being. They declare that through the medium of beautiful things they come into an intimate fellowship with this Being, and that from this fellowship and sense of harmonious relationship they are helped in their upward struggle towards the fuller life of man. Some carry their thought further and describe this fellowship as an immediate awareness of God. They say that under the stimulus of beauty they have a vision of God more vivid than can be obtained through the other media of man's contact with the Divine.

Let us hear from one or two their expression of this relationship.

William Drummond of Hawthornden, for instance, says:

Of this fair volume which we World do name
If we the sheets and leaves could turn with care,
Of him who it corrects, and did if frame,
We clear might read the art and wisdom rare:

[1] *The Story of My Heart.*

Find out his powers which wildest powers doth tame,
His providence extending everywhere,
His justice which proud rebels doth not spare,
In every page, no period of the same.
But silly we, like foolish children, rest
Well pleased with colour'd vellum, leaves of gold,
Fair dangling ribbands, leaving what is best,
On the great writer's sense ne'er taking hold;
Or if by chance we stay our minds on aught,
It is some picture on the margin wrought.[1]

William Wordsworth goes rather more deeply in the following well-known passage:

For I have learned
To look on nature, not as in the hour
Of thoughtless youth; but hearing oftentimes
The still, sad music of humanity,
Nor harsh nor grating, though of ample power
To chasten and subdue. And I have felt
A presence that disturbs me with the joy
Of elevated thoughts; a sense sublime
Of something far more deeply interfused
Whose dwelling is the light of setting suns,
And the round ocean, and the living air,
And the blue sky, and in the mind of man:
A motion and a spirit, that impels
All thinking things, all objects of all thought,
And rolls through all things. Therefore am I still
A lover of the meadows and the woods
And mountain; and of all that we behold
From this green earth; of all the mighty world
Of eye and ear,—both what they half create,
And what perceive; well pleased to recognise
In Nature, and the language of the sense
The anchor of my purest thoughts, the nurse,
The guide, the guardian of my heart, and soul
Of all my moral being.[2]

[1] *The Lessons of Nature.* [2] *Tintern Abbey.*

I take my third illustration from Michael Fairless:

Shut in by the intangible dark, we are brought up against those worlds within worlds blotted out by our concrete daily life. The working of the great microcosm at which we peer dimly through the little window of science; the wonderful, breathing earth; the pulsing, throbbing sap; the growing fragrance shut in the calyx of to-morrow's flower; the heart-beat of a sleeping world that we dream that we know; and around, above, and interpenetrating all, the world of dreams, of angels, and of spirits. It was this world which Jacob saw on the first night of his exile, and again when he wrestled in Peniel until the break of day. It was this world which Elisha saw with open eyes; which Job knew when darkness fell on him; which Ezekiel gazed into from his place among the captives; which Daniel beheld as he stood alone by the great river, the river Hiddekel. For the moment we have left behind the realm of question and explanation, of power over matter and the exercise of bodily faculties; and passed into the darkness alight with visions we cannot see, into silence alive with voices we cannot hear. Like helpless men we set our all on the one thing left us, and lift our hearts, knowing that we are but a mere speck among a myriad worlds, yet greater than the sum of them; having our roots in the dark places of the earth, but our branches in the sweet airs of heaven. . . . But surely the great mystics, with all their insight and heavenly love, fell short when they sought freedom in complete separateness from creation instead of in perfect unity with it. The Greeks knew better when they flung Ariadne's crown among the stars, and wrote Demeter's grief on a barren earth, and Persephone's joy in the fruitful field. For the earth is gathered up in man; he is the whole which is greater than the sum of its parts.[1]

And again, in another essay in the same volume:

I feel not so much desire for the beauty to come, as a great longing to open my eyes a little wider during the

[1] From "Out of the Shadow," in *The Roadmender*.

time which remains to me in this beautiful world of God's making, where each moment tells its own tale of active, progressive life in which there is no undoing. . . . I have lost my voracious appetite for books; their language is less plain than scent and song and the wind in the trees; and for me the clue to the next world lies in the wisdom of earth rather than in the learning of men. . . . I can never remember the time when I did not love her, this mother of mine with her wonderful garments and ordered loveliness, her tender care and patient bearing of man's burden. In the earliest days of my lonely childhood I used to lie chin on hand amid the milkmaids, red sorrel, and heavy spear-grass listening to her many voices, and above all to the voice of the little brook which ran through the meadows where I used to play: I think it has run through my whole life also, to lose itself at last, not in the great sea, but in the river that maketh glad the City of God. Valley and plain, mountain and fruitful field; the lark's song and the speedwell in the grass; surely a man need not sigh for greater loveliness until he has read something more of this living letter, and knelt before that earth of which he is the only confusion.[1]

Ruskin, who was filled with this passion for the beautiful, saw in the beauty of Nature a problem which demanded a solution, and the solution, to his idea, could only be found in a spiritual interpretation. The Æsthetic Emotion is a stirring of the Divine Spirit within us, and, as such, is a revelation of the God who is all-beauty, as well as all-truth and all-love.

And now, in writing beneath the cloudless peace of the snows of Chamouni, what must be really the final words of the book which their beauty inspired and their strength

[1] From "At the White Gate," in *The Roadmender.*

guided, I am able, with yet happier and calmer heart than heretofore, to enforce its simplest assurance of Faith, that the knowledge of what is beautiful leads on, and is the first step to the knowledge of the things which are lovely and of good report; and that the laws, the life, and the joy of Beauty, in the material world of God, are as eternal and sacred parts of His creation as, in the world of spirits, virtue; and in the world of angels, praise.

From Ruskin's virile faith, we turn to the underlying hope and deep desire for life's fullness, which was the keynote of the message of Richard Jefferies. We cannot discuss here the special problems of his life—the influence of poverty, disease, and despair upon his writings—suffice it that he was not only a consummate artist in his perception of the beauties of Nature, and in his power of word-painting, but he was also a searcher for the meaning of the beauty. He attempted in many of his finest essays, and in his wonderful spiritual autobiography, to see through the beauty to the Power within moving towards the perfection of life, and if he did not recognise this Power as a Loving Father, let us remember his early death and the sad circumstances of his life.

When I look in the glass I see that every line in my face means pessimism; but in spite of my face—that is my experience—I remain an optimist. Time with an unsteady hand has etched thin crooked lines, and, deepening the hollows, has cast the original expression into shadow. Pain and sorrow flow over us with little ceasing, as the sea-hoofs beat on the beach. Let us not look at ourselves, but onwards, and take strength from the leaf and the signs of the field. He is indeed despicable who cannot look onwards to the ideal life of man. Not to do so is to deny our birthright of mind.[1]

I cannot leave it; I must stay under the old tree in the midst of the long grass, the luxury of the leaves, and the song in the very air. I seem as if I could feel all the glowing life the sunshine gives and the south wind calls to being. The endless grass, the endless leaves, the immense strength of the oak expanding, the unalloyed joy of finch and blackbird; from all of them I receive a little. Each gives me something of the pure joy they gather for themselves. In the blackbird's melody one note is mine; in the dance of the leaf-shadows the formed maze is for me, though the motion is theirs; the flowers, with a thousand faces have collected the kisses of the morning. Feeling with them, I receive some, at least, of their fullness of life. Never could I have enough; never stay long enough. . . . The exceeding beauty of the earth, in her splendour of life, yields a new thought with every petal. The hours when the mind is absorbed by beauty are the only hours when we really live, so that the longer we can stay among these things so much the more is snatched from inevitable Time. . . . These are the only hours that are not wasted—these hours that absorb the soul and fill it with beauty. This is real life, and all else is illusion, or mere endurance. Does this reverie of flowers and waterfall and song form an ideal, a human ideal, in the mind? It does; much the same ideal that Phidias sculptured of man and woman filled with a godlike sense of the violet fields of Greece, beautiful beyond thought, calm as my turtle-dove before the lurid lightning of the unknown. To be beautiful and to be calm, without mental fear, is the ideal of Nature. If I cannot achieve it, at least I can think it.[2]

I turned to the blue heaven over, gazing into its depth, inhaling its exquisite colour and sweetness. The rich blue of the unattainable flower of the sky drew my soul towards it, and there it rested, for pure colour is rest of

[1] "The Pageant of Summer," in *The Life of the Fields.*
[2] Ibid.

heart. By all these I prayed; I felt an emotion of the soul beyond all definition; prayer is a puny thing to it, and the word is a rude sign to the feeling, but I know no other. . . Let my soul become enlarged; I am not enough; I am little and contemptible. I desire a greatness of soul, an irradiance of mind, a deeper insight, a broader hope. Give me power of soul, so that I may actually effect by its will that which I strive for. . . . I desired that I might do or find something to exalt the soul, something to enable it to live its own life, a more powerful existence now. I desired to be able to do something for the flesh, to make a discovery, or perfect a method by which the fleshly body might enjoy more pleasure, longer life, and suffer less pain.[1]

But is there not ugliness as well as beauty? It is best to answer this question by considering why some things in Nature do not attract us and do not give us the inward emotion which we have been considering. Our prejudice against the slackness of parasitism may well blind us to the wonderful fitness of the parasite to its life of ease. We may have a personal dislike because of our warmbloodedness to the cold sliminess of earthworm and slug, and ignorance and fear may prevent us from recognising beauty in insect and snake. Again, sometimes an unpleasant association may prevent us from appreciating the beauty of form or colour or adaptation. I can well understand what my friend and teacher, the late Professor Miall, meant when he wrote in reference to the love of mountains, "But a solemn feeling now mingles with my recollections of the innocent little peaks of our familiar Lake-country. On

[1] *The Story of My Heart.*

December 31, 1893, in the height of his manly strength and in the full enjoyment of his great gifts as an investigator and a teacher, Arthur Milnes Marshall fell from near the summit of Scawfell, and perished in a moment," for I have much the same feeling with regard to the wave-dashed coast of North Cornwall, since witnessing the tragic death of a friend, by a chance fall.

We must also rule out from our consideration the plants and animals that have come under the moulding power of man's selection.

When we have done this, and when we consider the healthy plants and animals in their natural environment, we shall find that the term "ugly" has almost, if not completely, disappeared from our lips; and our keener insight supports the reality of beauty in the animate.

The appreciation of beauty is due to an inward instinct or intuition, and we become aware of it as it rises up into the mind, and by the attempt of the mind to express it in some way. It is first of all a stirring of the spirit, and this stirring becomes recognised or realised by the mind.

These ideas link on the sense or feeling of beauty to the moral sense, the feeling of right and wrong. We are really dealing with another aspect or another activity of the same spirit. In a previous chapter we have attempted to see how the moral nature of man arose amid the evolutionary environment, and have tried to understand, in some measure, the causes which brought this about. Here we are faced with a similar problem, but, although some hint of the evolution of the æsthetic emotions can be traced in the sex emotions of lower animals, I do not propose to discuss this here. The idea of beauty, of course, belongs to the stage of self-consciousness, but we can say this much about it, that the awareness of beauty has its double root in the environment of beauty and the existence of the inward spirit which can respond to this outward beauty. Just as man, as he evolved into a moral agent and became aware of the moral idea by living in the environment of freedom and endeavour, with the control brought about by social life; so man has, at last, become aware of the beauty in this same environment. The complete environment has been there around him from the beginning. The beauty of the world did not begin when man first became aware of it, any more than the air environment when the pioneer fish-amphibian first became aware of the fact of air. This new fact in its environment caused profound changes in its structure, as we have seen; the fact of beauty has also had, and is still having, a profound effect upon the personality of man. Beauty, then, is a part of the scheme, otherwise man would not have been able to realise it. The fact of the pervasiveness of beauty, and the fact of man's relation to it, go to show that we are not dealing with a gigantic illusion.

Man's æsthetic activity is part of his birthright as a man, and the exercise of this activity brings its own satisfaction; and the satisfaction only comes if the activity is exercise for its own sake.

We have been careful in our discussion to think of beauty in relation to Nature and not to works of art, which show the emotion stirred by beauty expressed through the technique of an artist. Of course works of art—whether in poetry, in painting or in music—have power to awaken the activity of this æsthetic emotion, and we owe our increased power of enjoyment to the training in insight obtained from such prophets of the beautiful. When we appreciate works of art we are re-creating for ourselves, in some measure, the intuition of the artist. It is only thus that the works of art can have any æsthetic value for us. What about this act of re-creation when we admire the glories of the sunset, or the brilliant Ampelopsis on the gable-end? Are we re-creating the intuition of the Divine Artist? In contemplating with a sympathetic insight the beauties of Nature, are we in touch with the art of God? This line of thought is well expressed in the following passage from S. A. McDowall's study of the subject:

The unsatisfyingness of beauty s due to the fact that you are taking and not giving. In order to give *something*, to others, though not to the object that roused in you the sense of beauty, you create by some technique. What is it you are receiving? An intuition, which you express to yourself creatively and to others through its effect on your character;—to which further, if you are an artist, you give external, technical expression. This intuition which you receive is the first stage of knowledge—of the knowledge of Reality. So far, agreeing with Croce, we agree with Bergson; and moreover we leave room for mysticism, since mysticism becomes the appreciation of relationship, and logic paves the way for suitable activity to develop our side of the relationship. The meaning of this becomes clearer when we consider Croce's explanation of the process of perceiving beauty in the work of an artist—be it picture, symphony, or poem. He points out that in appreciating a work of art you enter into the mind of the creator, follow his intuition, and create the expression afresh for yourself. On the degree in which you can do this depends the fullness of your appreciation of the work.

But when you see beauty in a natural object the matter is less clear. Croce would say that you are in the first stage of knowing that object, and he is unquestionably right so far. But can we not, using the analogy of the picture or the poem, go on to say that you are following out the idea of the creator of the natural object—that you are in touch with the Cosmic Idea, which is the Idea of a Personal God ? If so, there is indeed room for mysticism, for mysticism becomes simply the realisation that you are in fact doing this. Moreover, Beauty and Love at once fall into relation. Beauty is not simply expression, but the expression of a relation, and it is incomplete because the relation is not yet reciprocal. Love is that relation itself.[1]

It is along such lines as have been indicated in this chapter that, I believe, we may find the way to an understanding of beauty. To find beauty in Nature—an inherent part of the scheme —serves to strengthen our belief in the Beauty of God, and this belief in the Beauty of God will surely help us to hold fast to the faith that finds goodness in the same scheme.

[1] *Beauty and the Beast*, p. 31.

BOOKS

BOSANQUET, BERNARD. Three Lectures on Æsthetics.
History of Æsthetic.
LEE, VERNON . . . The Beautiful.
The works of R. JEFFERIES, THOREAU, W. H. HUDSON, etc.
The poetry of SHELLEY, KEATS, TENNYSON, etc.

CHAPTER VII

THE KINGDOM OF MAN AND THE KINGDOM OF GOD

" There's the Sperrit o' God in all things and all times, weekdays as well as Sunday, and in the great works and inventions and i' the figuring and mechanics. And God helps us with our head pieces and our hands as well as with our souls." *Adam Bede.*

" It is not the Universe as seen by man's bodily eye that is divine, but the Universe seen as it really is, seen as God sees it, seen, in the unity and totality of its all-pervading life by the all-seeing eye of the all-sustaining soul. It is not until we can see the Universe as it really is that we are free to identify it with God."

The Creed of Christ.

" When I came into the silent assemblies of God's people, I felt a secret power among them, which touched my heart; and as I gave way unto it I found the evil weakening in me and the good raised up."

ROBERT BARCLAY.

WE have endeavoured to understand the facts of man's ancestry; we have tried to make it clear that there is no shred of doubt about his place in Nature; he is one with the scheme—at the same time both its product and the expression of its purpose. All the same man is unique, and it is no play upon words to recognise a *Regnum hominis.* In recognising this kingdom it will be clear, from what has gone before, that there is no unbridgable gulf twixt man and the beasts, for his entry into his kingdom has been along the evolutionary pathway. Although every step of the journey cannot be mapped out, we can see sufficient of the pathway to have no doubt of the method.

What do we mean then in saying man is unique? We think, first of all, of structural peculiarities, which, though few, are important. The chief ones to note relate to his upright carriage and the changes brought about thereby, such as the freeing of arm and hand for grasping and for making things, and to the large brain. We should next think of the environmental peculiarities without which these structural changes would have had but little chance of survival, and see in the social life and the prolonged infancy important factors in his making. Some biologists put these first as of chief importance; but it is difficult to speak of first and second where all are interwoven into the web of intimate relationship. Brain, social life, infancy and a grasping hand are all interdependent.

Leaving the details of structure and of environment, we should next consider the way in which, by the use of his intellectual power, man has obtained control over Nature. Perhaps it would be better to say has been obtaining control, for he has not yet learnt all his lessons, nor does he apply faithfully those which he has already learnt. Man is certainly unique in that he is aware of his own position in the scheme, and in that he attempts to guide his life according to some plan or ideal. In this fact of awareness, we begin to step out of the animal phase into the mental and spiritual phase. In this fact of awareness we have become conscious of a kingdom to be possessed, and, in going forward to take possession, we have come into hearing distance of new voices, and, as we have listened to and obeyed these voices, we recognise a new guidance in our lives. In this fact of awareness we find that there is a *regnum hominis*, but every step forward into this promised land makes clearer the fact of a still greater kingdom awaiting possession, and that this—the *Regnum Dei*—is the real end and purpose of the whole.

The uniqueness of man, then, is seen to be a fuller appreciation of the environment, both the material and spiritual aspects of it. By becoming aware of his place, and by virtue of his intelligence and resourcefulness, he has been able to combat many of the natural legacies of his animal ancestry, and so to become, in a large measure, the master of his fate. Above all we see his power to direct his life according to some idea or ideal; this, the last of his acquisitions, is the greatest of all, and we can quote with approval Professor Weismann's remark: "Spirit is the deciding factor." Let us consider more fully what is meant by man's control over Nature. We mean first of all that man is able to select, to a large extent, the kind of environment in which he will live. This means that he can have restrictive or expansive conditions around him. The nurtural pressures are under his control. Of course, we do not mean that every individual man has this selective control, for men are still held down in slavery by economic conditions; these economic conditions, however, are man-made, and can be altered by man if and when he cares to do so. Man is a child of Nature in his inheritable qualities. He is subject to variation as all other organisms are subject to these strange new-comings, indeed, there is additional chance of variation in the very complexity of his structure; and these variations, along with all the other features of man's inheritance, come into relationship with the nurtural pressures. It is quite clear, from experiment and observation, that what happens in the case of lower animals also happens here. If the nurtural influences are abundant and suitable, the various features of the child's inheritance can expand and flourish. The question whether any faculty will come into its full expression depends largely upon the environment, that is upon the nurtural conditions surrounding the child. In the absence of a suitable environment, qualities which were there in a dormant condition remain undeveloped and unsuspected. To take but one instance of this, the changes in the method of education, and the widening of its content, have revealed in many children characteristics of a surprising interest, characteristics which have been unsuspected until a favourable nurture has awakened them into outward expression. Education works no miracles —it simply allows the development of the child's normal inheritance, and gives scope for his creative power so often repressed by the environmental pressures. What is true of education in its narrow sense is true of education in its wide aspects.

We must be careful to think of "nurture" in as wide a way as possible, for it includes all the liberating forces which play upon human life as well as all the imprisoning forces. The control of man follows from his realisation of the power in this nurture; for this nurture can encourage or discourage, it can supply the needs for healthy, holy (i.e. complete) development, or it can withhold these necessary things, and by starvation—physical, mental and spiritual—retard the growth and development. It is not clear yet what permanent effect it has upon the nature. At first sight it looks as though the moulding effects of the environment would be transmitted direct to the offspring. This is not so, and considering man's repeated failures, misunderstandings and feeble control arising from his ignorance, it is just as well: we could not have the good passed on without the bad. Although there is but little evidence of an immediate impression on the germ-plasm made by the nurtural pressures and circumstances, yet the thesis of germinal isolation is no longer tenable, and the close inter-relation of the sexual organs and their products with the rest of the body is clearly recognised. The fact that the germ-cells are formed in the embryonic period removes the possibility of striking changes becoming impressed upon them; but we know too little of the microcosmic germ-cell to rule out the possibility of any change. We do not know how or why the variations arise, though probably one of the factors will be the changing environmental stimuli transmitted through the body to the germ-cells. Some biologists declare, and there seems to be good reason for their faith, that if nurtural pressures or encouragements are applied generation after generation, then some permanent effect can be detected, and the repeated influence will make its mark upon the germ-plasm. Whether we agree to the transmission of acquired characters in this modified form or in no form, we cannot fail to recognise that the deciding factor from both points of view is the nurture. Even if there is no permanent effect upon the nature, the gains of one generation can be recaptured in the next by the right arrangement of nurture.

It may be that the very persistence of these lower characteristics is necessary in order to provide material for individual conquest. Although the far-reaching effects of these strands of our animal ancestry may be modified by the collective mastery of associated life, yet the stage is set anew for each individual, and each individual has to win through and triumph over the enemy for himself, albeit with all the aids ready for his use. The story of the bodily recapitulation of racial stages during ontogeny seems to support this view, and to suggest that there is a similar recapitulatory conquest in the spiritual life, and that the very freedom won in the scheme has value only for use in the conflict with and transmuting of antagonistic forces.

The facts of social life and of the impossibility of living in isolation bring in new factors. The sifting by Natural Selection is largely interfered with by man. Life in societies allows a much larger range of individuality, and the tendency of association will be to protect individuals from the rigors of the old system. In any animal community there is a lessening of mutual struggle; but in the communities of man the process has gone still farther, for it comes under the control of intellect and religion. Life is governed under the sanctions of a code handed down from one generation to another. There is also the possibility of a social variation making possible the alteration of the whole course of social life. These strange new turns in the path of progress—new patterns in the fabric of human society, come up before the court of social selection for sanction or dismissal. Man is still a long way from a real control even of those features of his life which by this time should have yielded their secret to him. In spite of wonderful advances in the control of disease, for instance, ignorance still holds man back from full possession of his kingdom; and there are many shadows which can be dispelled by a more faithful development of his life along the pathways revealed by science. Wrong turns have been made and these have brought grave results, but even these can be rectified if man will face in a right spirit the problem of the individual in relation to society, and in relation to the end towards which the whole scheme of life is clearly meant to move. Social evils which show themselves in poverty and in war are man-made and can be man-cured.

These evils are due to the failure to understand the message of evolution resulting in a failure in method. Man has in his hands the power to control his life, and he has failed in the use of this weapon forged through the long ages of the past. He has used it for selfish and lesser ends. He has now to use it for unselfish and greater ends. The widening of the social units, the breaking down of the barriers between small groups entrenched within the fastnesses of the larger groups, the spread of real sympathy and love; these are some of the obvious underlying ideas necessary to the right social use of the controllability of life.

But the real way forward to the establishment of the right relations in his kingdom is for man to use the method of religion as the basis for individual and social life. What I mean by this can be best expressed by a definition of religion quoted by Professor Ray Lankester in his address on "Nature's Insurgent Son,"[1] from the *Life of Bishop Creighton*. "Religion means the knowledge of our destiny, and of the means of fulfilling it." This is what I mean when I say that the method of religion is the way for man. The great characteristic of man is his power of seeing in the system of Nature a scheme in which he finds the roots of what within him calls upward to a better, more beautiful and more complete life. He knows himself as a moral agent, but he knows, too, that his moral nature has grown out of the ground of the Universe. He feels within himself the emotional thrills as he contemplates the beauty

[1] *The Kingdom of Man.*

around him ; but he is conscious at the same time that the beauty is inherent in the very substance of the Universe. He knows himself as a reasonable soul, but he feels the threads of the fabric of his life reaching back into the dim past. These feelings of continuity are supported by all that science can fathom of the evolutionary story and of the life of man. Professor Thomson puts these thoughts into beautiful language in the last of his *Gifford Lectures.*

> From the intrinsic order and intelligibility of Nature, which the rise of the magnificent scientific edifice proves, we may not be logically permitted to make a transcendent inference to an Omniscient Creator, but it is in that way the heart of Man points. Our belief is that the Logos is at the core of our system, implicit in the nebula, as now in the dew-drop. It slept for the most part through the evolution of plants and coral-like animals, whose dream-smiles are a joy for ever. It slept as the child sleeps before birth. It became more and more awake among the higher animals,—feeling and knowing and willing. It became articulate in self-conscious Man,—and not least in his science.[1]

We have seen something of this order, which spells intelligence and therefore a mind moving towards Reality ; we have discussed the beauty of Nature, and the satisfaction and uplift this gives to the spirit of man, and we have traced, in some measure, the lines along which the movement towards goodness has proceeded. In seeing these things, we become conscious of our destiny, we find within ourselves faith in a fuller life, and

[1] *The System of Animate Nature*, p. 637.

a belief in goodness and beauty as constructive forces, with evil and ugliness recognised as destructive forces. Such a faith and such a belief drive us along the pathway to knowledge and reality ; not as a vague dream or illusion, but as the working out of a wonderful purpose.

If a purpose, surely a Purposer. Can the Purposer be other than God ? To quote Professor Thomson again :

> Scientific reconstructions are not arbitrary projections, for they work. In this sense there is rationality in Nature. But if there is rationality in Nature, must we not go farther ? For, as Aliotta has put it, " he who believes in the objective value of his science must then also believe in God. If an absolute thought does not exist, Nature cannot be rational." [1]

Cannot we begin to see that God has been present all through the great drama ; or to use another metaphor—all very imperfect in their suggestiveness—cannot we see God as " the ground—the continuous ground of the world's being ? " I was attracted to this metaphor when reading *Reconciliation and Reality*, by Professor W. Fearon Halliday.[2] In the chapter on Providence he refers to the inadequacy of saying that God is the Great First Cause, and introduces this idea of ground as a " wider and deeper one than that of cause." I have since come across the same idea in that interesting collection of mystical visions entitled *Revelations of Divine Love* ascribed

1 *The System of Animate Nature*, p. 637.
2 Vol. ii. in The Christian Revolution Series.

to Lady Julian of Norwich, who lived during part of the fourteenth and fifteenth centuries. "God is nearer to us than our own soul." God is "the ground; He is the substance, He is the teaching, He is the teacher, He is the end, He is the meed for which every soul travaileth."

We have spoken of God in these chapters sometimes as the Perfect Environment, but this is obviously open to objection; God does indeed permeate all life as well as being the ground and substance from which the life draws its supplies for growth. He is more than this as the author is more than the print—the code of his message. God then is Artist-Creator. Man enters into his kingdom only to find that he is called to live in a still more wonderful kingdom; and by the very act of trying to live in this greater kingdom he is creating it or rather he is assisting God in the creation of His Spiritual Kingdom. The creation of this Kingdom of God is the purpose of the whole scheme.

This Kingdom has been given various titles, none is really satisfactory, but I prefer Josiah Royce's names "The Beloved Community" or "The Divine Community" better than the usual one, for they do not suggest the Kingship of God to those to whom the title of King seems a little out of date. For the kingdom is a loving fellowship, and this fellowship of Divine Love is at once the end of all this vast struggle and the explanation of it. Herein method and motive become harmonised.

In our review of the evolutionary method and

in our explanation of the working out of the scheme of life, leading through freedom to love, have we thrown over the Providence of God? Can God interfere with the normal working out of His purpose? Is His arm shortened that it cannot save? Is prayer banished? Has God been removed to a distance? Do we regard Him as a spectator, watching from without, powerless to interfere? Is it not true that in solving some problems and in clearing away some doubts we have raised greater difficulties than were there before? At first sight it may seem that these questions do raise insoluble difficulties; but as we look more closely, and as our eyes become accustomed to the light revealed by Science, we shall be ready to give a deeper meaning to all these things. No, the Providence of God is very sure, prayer becomes more real, and God is very near to every human soul.[1] As a result of our enquiry we see deeper into the mysteries of God and see His hand in every happening. "When the half-gods go, the gods arrive," wrote the seer of Concord, and that is the hope that carries us forward throughout this quest. To have a more reasonable faith, to establish a purer trust, to find a more satisfying idea of God and of His way for man, these things form the aim of this enquiry. Above all, to see the revelation of Jesus against the background of the evolutionary story, with its winning of freedom by man, is to bring a new and deeper meaning to the Gospel story.

[1] For a fuller discussion of Providence, see pp. 157–166.

God is not an outside God; He is, as we have said, " the ground of all our being " ; and the fact that man is conscious of the Spirit of God within him, leads him to believe in the presence of the Spirit of God in all things. This spirit is that mysterious somewhat in life—not added on or superimposed, not even an influx—but something inherent in the very stuff of life.

Religious philosophers speak of the Immanence of God, and the biological philosophers speak of animals as psycho-physical beings. Are not these different aspects of the same idea? If this is a true statement, and I am profoundly convinced that it is, then God is never absent from His world, or from any part of it.

Most people in speaking about God, think of Him as the great Personality above and outside the Universe and they think of Him in terms of perfect holiness, beauty and reality. In philosophical language this is called the Transcendence of God, and to the biologist the terms Purposer and Creator convey a more understandable idea. It is abundantly clear, however, that both ideas are aspects of the one God. The Transcendence and the Immanence are aspects of the one supreme being who is the author and the sustainer of all life, and who has been drawing man towards and into this divine and loving fellowship. It may be well to give here a few paragraphs from Professor W. Fearon Halliday's *Reconciliation and Reality*, in which he deals with the doctrine of God from these two aspects, for his treatment of the problem is very suggestive and helpful.

It will be found that most of the difficulties in connection with the doctrine of God have arisen because we have thought of Him in spatial terms. It has long been a commonplace in theology that God is Transcendent and Immanent; but both these terms are *spatial.* To say that God is Transcendent is to say that God is outside the world, but not identified with it. But if we speak of the Universe, " Transcendent " could only mean that He was a part of the Universe, but not identified with our part. From this conception derive both the Epicurean thought of God as of One in a far-off tranquil world and the more modern notion of Deism.

The doctrine of Divine Immanence on the other hand is well illustrated in the Stoic thought of God as the soul of the world. These systems of Immanence proclaim a diffused, depersonalised God regarded as life or more abstractedly as Reason. The result is Pantheism with no distinction between God and the world. In both cases there is an inevitable tendency to think of God materially. The two doctrines leave us in the dilemma, Is God a part of the Universe? then He is limited; or is God the whole Universe? then we have lost Him as He is identical with us. But to religion the thought that God is in the world is just as necessary as that He is above the world. Now, as has been said, Transcendence and Immanence are spatial terms; . . . space is here taken to be the relation and form under which we apprehend matter. To apprehend God in terms of space therefore is to apprehend Him in terms of that which is related to the material, in terms that is, of what is phenomenal and not real. It will help to a solution of this difficulty if it be realised that at the same time we are living in a world that is seen and that also is a world that is unseen—our bodies are in space, but *we* are not in space. It is because we think of one another as in the body, and therefore in space, that the confusion arises. These two worlds—the spatial and the spiritual—are not only distinct, but have contrary characteristics. It is impossible to take a lump of matter from one place and put it in another without creating a vacuum and filling a vacuum. But it is far different in the spiritual world. A man communicates his thought without losing it; his friend may receive it and assimilate it and both are richer, So it is with love. The reason is that thought and love are not spatial. . . . The fact is that men have never seen one another but only one another's faces; the face is the medium through which the self is expressed; time changes its every atom, but the personality which is manifested through the face retains its identity. All the noblest things in life are connected with this unseen and spiritual world, loyalty and love and truth. The importance of events and deeds is not in themselves but in the meaning which we find in them, and the value which they have for personality. But meaning and value are not spatial terms. Terms, therefore, such as Transcendence and Immanence may be used only symbolically or metaphorically of that which is spiritual and personal. Now, strangely enough, there is nothing known to us which can be called at the same time Transcendent and Immanent except personality. In friendship men are transcendent from one another because each is self-conscious; they are immanent in one another through love and understanding. But in neither case is the relationship spatial. If, then, we cannot do otherwise than think of God as both Transcendent and Immanent, we have a key to the meaning of this, not through the outer world of nature, but through our own personalities and experience.[1]

We can only know God in His relation to us, and we can think of Him under the dual name of Creator-Father. All that we know of goodness, truth and beauty in human personality will also be found in Him the Perfect and Complete Personality. This argument from man to God gives a new meaning to the revelation of Jesus,

[1] Pages 46–49.

in whom we see God revealed through the medium of perfect humanity.

We are conscious, too, of the creative activity, which is a part of our personality, and we know that love is the expression of our highest and best relation to other people. So when we try to understand the personality of God and think of Him as perfect Love, we imply at once the need of some objects, of some persons rather, for the exercise of His love. If we cannot think of love in the abstract, apart from fellowship and its exercise among persons when we are considering the love of men, surely it is legitimate to think that the same is true of the love of God At least in using the terms "love" and "father" Jesus revealed to us something of this aspect of God; and His revelation, indeed His whole life, depended upon the fact of loving relationship between Himself and God.

We can therefore think of the Immanence of God as the influence or spirit of His Transcendence, which, poured out lavishly, has been the inward urge throughout this long story of man's becoming, and has been the real cause of his advent. We can look upon this as the great creative act of God—the creation of persons able to appreciate and return the love which has been so wonderfully given to men. But the creative act was not one great action done once and for all, the method of evolution, which we have been trying to understand, needed the continual process of creation. The spirit of God is Immanent, and the process has been a working out towards man—man free and creative, otherwise the creation for the purpose of loving fellowship would have failed. The purpose has not failed. Man knows himself called to be a son of God, and he has within him the longing which cannot be satisfied until he finds in the love of God the answer to all his difficulties and the meaning of all his strivings.

The story of Animate Nature is one long evolution and gradual unfolding—exfoliation is Edward Carpenter's word—of the essential, inherent spirituality. We have tried to describe the working of the method, action-reaction, interaction, first a little better vehicle and then this reacting upon the appreciation of the environment and so making possible a development of the psyche. This slow experimental method is necessary, seeing that goodness and love were the aim, and the Goodness of God could only have such an aim. This makes us examine traditional views of God's Providence, and to reject those that are based upon the idea of a spasmodic or arbitrary interference in the process. God's Love has been the cause of and the power within the age-long drama of the world's becoming, and we cannot expect that His Providence would work across the scheme, defeating the object of His creative activity. In demanding that God should, as a special favour, alter the regular working of the cosmical laws, we are asking Him to contradict His nature; but we are not asking Him to contradict His nature or the scheme of man's freedom when we pray for His spirit to enter and fill our lives with conquering power. For God to use His power to compel men to act in such and such a way would only be possible in a mechanical world peopled by machines. In such a world love would be unthinkable; and machines can neither be reasonable beings nor moral agents. Men are not machines, and the world is not a huge mechanism: no, men are akin to God; there is no great gulf fixed between human nature and Divine nature, and God's love is everywhere. "If God be indeed the end of all existence, He must needs fill all things with His being. If God is love, His arms are round the entire universe, and there is no creature anywhere unloved by Him."[1]

Men, then, live within the influence of the love of God, but it is not forced upon them; rather it is the nurture or environment of the soul, and men can freely respond to it; or they can as freely reject it, though their rejection will limit their progress towards fuller freedom.

It may be well to gather up here the ideas upon the Providence of God which have been implied and suggested throughout our enquiry, for the problem of God's Providence is vital. Unless we have a right understanding of the relation between God and this world of evolving purpose we shall have a false idea of God, and a false idea of God will prevent us from entering into that fuller fellowship with God, from which comes the power to go forward as to victory in our conflict with our difficulties and evil tendencies. How does God deal with men? Does He deal directly with them, and can He override the human spirit

[1] Prince Troubetzkoy.

by an arbitrary exercise of power? Can God interfere with the normal working of the laws of nature—His own laws? Can He set them on one side for the sake of an individual life?

Can God really do anything, or is His action limited? Does God deal with men as with individuals, or does the evolution-idea destroy all idea of God's personal love and care? Is God responsible for pain and evil? Why does He allow the good to suffer? Does God really care for *me*, or is His care for the race? Is there any value in communion and prayer apart from the mental suggestion and stimulus experienced by the individual?

The questions are interminable, and the ones suggested here show how deep the question of Providence goes and how important it is to attempt to give some answer.

Many of these questions have been answered directly or indirectly, and what follows is an attempt to bring together a few general ideas upon the Providence of God.

We probably make our first mistake by imagining what we might do if we had God's power. We are so domineering and so prone to use any power and influence for personal advantage and advancement, that we often credit God with something of the same motive. If so, we are thinking of God as an oriental despot to be propitiated, or cajoled into giving gifts to favourites.

The first thing we have endeavoured to establish is that God works naturally. The so-called laws of nature are a sign of the Divine order, and we have tried to see how necessary this order in the scheme is for the development of man. The scheme is a sign of God's presence, an expression of His mind, even though in the lower circles of activity there is no conscious realisation of His presence. Just as in the story of man's religious development we recognise that many are living in unconscious fellowship with God, as shown by their striving towards the good, the true and the beautiful, although they do not recognise the meaning behind their striving; so in the ages before man's rise into self-consciousness the spirit of God was there and the love of God was there behind all the upward struggle and endeavour, even though there could be no recognition of it. The proof is the same in both cases. We judge a man's life by the results, by the evidence of the repeated and progressive victory of the good, and we judge the scheme by its result in the evolution of man. Both fruits speak to us of the Providence of God. If we may use the parallelism again, we could tell of the acquisition of new power, when a man can relate his own strivings to the purposes of God, and find in a conscious fellowship a new environment of love and faith, in which he becomes a new creature, and see this same change and rise into a fuller life as the great step in the story of evolution. We can perhaps make the point clearer by using again the metaphor of a fish in its watery environment. To such a creature the existence of air was not appreciated even though it was utilising the air dissolved in the water for the vital respiratory processes. It became more known (of course not consciously) when the pioneer fish began to gulp in gaseous air into its mouth and swim-bladder. As a result all kinds of new adaptations came into use resulting in the fish becoming an amphibian and then the amphibian a reptile—animals living in and using more of a higher environment. It is only suggestive and the metaphor must not be pushed too far; but it may suggest how the all-embracing spirit of God has been waiting for man's conscious recognition and use, and that before this period its presence was necessary and vital to the working out of the scheme. We have also attempted to show how the conditions of social life made this recognition possible, and the fact that some feeble kind of morality is recognised by all intelligent social animals, helps us to realise the pervasiveness of God's spirit.

The orderliness of the scheme is a necessary feature of the evolutionary method. No school of experience is possible unless the happenings in the physical world can be relied on. The scheme leads to the development of personality —everything supports this idea of its purpose— and we know that one of the essentials of personality is free activity. This freedom has to be won by the conquest of conditions governing life. Animals are not determined from without by any controlling act of God, either sudden or continuous, but they determine their own life by the exercise of real though limited freedom. Thus although we believe God is omnipotent, He doesn't use His power to override either man or the lower animals, but uses His power and influence in such a way as to educate them. "May it not be that we shall not find less of God in life and not find His operation less adequate to our spiritual needs because we discover His method to be patient enough to pass round by way of persuasion and education through our errors and failures." [1]

If this is true for man, it must surely be true for the whole scheme. The facts of man's ascent from and solidarity with the lower animals is so certain, and the rise of his mental and spiritual characteristics so clearly natural, that we cannot but believe that this Divine method of patience and education began far back—indeed is implicit in the whole drama of evolution. The idea of Providence is thus extended and enlarged by making it universal in its action. The Providence of God has never been absent from the scheme, and our scientific enquiries instead of doing away with the idea have increased its scope and importance. All Nature is sacramental and God's presence is everywhere; nothing is apart from Him.

God does not conduct His rivers, like arrows, to the sea. The ruler and compass are only for finite mortals who labour, by taking thought, to overcome their limitations, and are not for the Infinite Mind. The expedition demanded by man's small power and short day produces the canal, but nature, with a beneficent and picturesque circumambulancy, the work of a more spacious and less precipitate mind, produces the river. Why should we assume that, in all the rest of His ways, He rejoices in the river, but in religion can use no adequate method save the canal? The defence of the infallible is the

[1] *Grace and Personality,* John Oman, p. 9.

defence of the canal against the river, of the channel blasted through the rock against the basin dug by an element which swerves at a pebble or a firmer clay. And the question is whether God ever does override the human spirit in that direct way, and whether we ought to conceive either of His spirit or of ours after a fashion that could make it possible. Would such irresistible might, as would save us from all error and compel us into right action, be in accord either with God's personality or with ours ? [1]

The dealings of God with man are, for most thoughtful people, their chief problem and their greatest difficulty ; and in giving the above quotation and in considering the ways of God in evolution, I wish to suggest that there is a unity about God's Providence, and that we must expect Him to exhibit the same principles of regularity and the same consideration for man's freedom. In support of this regularity one might urge the discoveries of psychology. Very much that was regarded as a special dealing of God with an individual, a sign of direct action, is now seen to be God working through natural and understandable psychological channels. Let us be very careful, however, not to think that to find an explanation in terms of psychology of some healing or of an accession of power for victory over sin, is to remove God from His Providential care. To find that God works indirectly along natural lines is not to deny the action of God. To give an explanation of the working of electricity does not deny its existence, nor does it explain what it is. To

[1] *Grace and Personality*, p. 8.

see a little more clearly how God influences man, does not do away with God or with His influence over man; it should make both more real, more reasonable, more certain. The whole purpose of this book is rather to emphasise the fact that God is seeking in every situation and by means of every element in our environment (within and without) to make known His will and His presence, and that He does this naturally and not arbitrarily.

The discoveries of psychology are therefore most important, for they reveal the method of God in dealing with man, and help us to believe more truly in the personality of God. The discoveries of psychology as to the influence between persons explains very much in religious experience which seemed abnormal and miraculous. They begin to show us the channels along which the love and power of God may come to us. Religious experience is not explained away by this illumination; religious experience will always be the chief witness to the power and love of God. The result of the discoveries of psychology is to make clear the naturalness of what has often been regarded as a special favour granted to certain persons, or a special grace obtainable through certain magical rites. Another great gain is the knowledge that the limitation of God's influence and Providence in our lives is really our fault. We are the limiting partner, for although God is longing to help us—indeed He has been waiting throughout the ages for this very chance—He is not going to force His influence upon us. Love, whether of a human or a Divine person, can only lead from before, it cannot drive man along the way of rightness. We get our best idea of Providence by considering how we influence one another. We know that there must be the willingness to enter into the loving fellowship with another person before we can receive or give the love which means so much in our lives. The Divine spirit is everywhere and always acting in the world, but it cannot act fully in our lives if we shut ourselves away from conscious fellowship with God as the Person who matters most in our lives. Our experience of the helpful influence of others will also prevent us from setting limits to what can be accomplished in us, and through us, when we come into that perfect fellowship. Religious experiences throughout the ages confirm our caution in setting limits to the transforming power of God's Holy Spirit.

These views about Providence are also confirmed by the life of Jesus, the greatest example of the power of God's spirit operating in a human being. Jesus showed how the Supreme Spirit was at work in all phases of life, and that if only men would open their hearts to the Divine influence, their lives would be flooded with tranforming power.

These views about Providence help us too, I believe, to understand how, by man's refusal to accept salvation, tragedy and error are so prevalent in human life. If only we could realise how God can help us in the uphill climb, and in the personal conquest of limitations and difficulties, we should not remain so long in the darkness of doubt and despair.

But does not this aspect of our problem suggest that God is not omnipotent? If we mean that there are things God will not or does not do, we answer "Yes" for He is, Himself, limited by His scheme. But a self-limitation in order to achieve His purpose is no argument against God's Omnipotence. To quote again from Professor W. F. Halliday:

> The end or purpose of life is only attained through God and man together; man cannot attain it without God, without recognition of the moral order of reality, but neither can it be attained until man makes God's purpose his own, and actively develops himself in its realisation. This does not in any sense involve God as finite; a God Who could not limit Himself to achieve His purpose would not be truly omnipotent. This self-limitation is necessary because of the nature of that purpose, which is the creation of a realm of persons or ends.[1]

But does not the scheme seem to work out towards the production in the future of the superman, with but little or no regard for the ordinary man or woman? Can there be any cognisance of those who fall by the way, or for those who have gone before, and to whom the fuller vision in Jesus was not vouchsafed? What of those cut off by disaster and by the sin, folly or ignorance of man? Their chance of personal development was lost through no fault of their own. What of the sinner and of those who reject the saving health of God?

[1] *Reconciliation and Reality*, p. 91.

Certainly the scheme fails and is a gigantic blunder unless it aims at the raising of all into this loving fellowship, and this it cannot do unless there is a survival and a continuance of personality after death. Such a survival will give the chance for progress and development. Only thus could one see how compensation could be given to each individual; and this compensation must take the form of further opportunities to make up for the lost opportunities, otherwise there could be no growth in personality. The belief in the continuance of opportunities for our education depends also upon faith in the character of God as Perfect Love. This vision of continuity goes further; it seems to suggest that there can be no private salvation either here or hereafter, and that neither we, without all the other struggling souls, nor they, without us, can be made perfect or can attain the final purpose. This is surely the thought of the inspired writer of the Epistle to the Hebrews, when after his fine summary of the faithful heroes of Hebrew history eager for the establishment of a better land he said:

> Although by their faith all these people won God's approval, none of them received the fulfilment of His great promise; for God had provided for them and us something better, so that apart from us they were not to attain to full blessedness.[1]

God's purpose is personality, and the development and enhancement of personality can only come about as the result of fellowship between

[1] Hebs. xi. 39, 40 (Weymouth's trans.).

our spirits and the Divine Spirit. The purpose is both far off, and here and now. It includes all men and women, and yet it concerns itself with each individual. The purpose depends upon the loving fellowship and mutual help between men, for, although God mediates His presence to us in many ways, one of the chief ways is through the love and kindness shown by man to man. We are called to co-operate with God, to be used by Him as instruments of His love, and because of this, the Providence of God is largely recognised in the providences of man to man. A story may illustrate this point. One who had lost faith in the presence of God came to an old woman friend of hers for help and guidance. After some discussion they knelt to pray, and the old woman said, "If you cannot believe God is here, ask Him to stretch out His hand and touch you." As they prayed the old woman put her hand on her friend's head. The troubled one rose from the prayer strengthened and comforted and said, "He did touch me, it was so wonderful, so loving, so kind, just like your hand!" "It was my hand," the old woman replied, "did you think God was going to make a long arm out of heaven and put it upon you? No. He just took my hand and used that."

We come now to think of the various ways in which the presence of God may become real and consciously known to us. There must be the desire to know the grace of God, for man can shut out from himself a fuller conception of God's goodness, by a refusal to use the methods which religious experience has proved of value, and which psychology is now showing to be natural and normal. Any circumstance may open the way for this personal realisation of the presence, for all life is sacramental and incarnational; but it is natural that some circumstances may arise which make the presence seem more real, and which enable the individual to get into that right relation to God whereby power flows into his life. To some people certain rites have the power of making real the presence of God, and the rite in most use is the Eucharist; but as a Quaker I wish to urge that there is a danger in putting forward one special rite or a group of rites. The danger is lest we imprison God within our sacramental walls and imagine that there are no other ways in which He normally appears to men. I do not wish to suggest that God is not found or more fully known in the practice of special sacraments; what I do maintain is that these special sacraments are not the only means of grace, nor are they the normal means of grace, and the elaborate claims of their special importance are not warranted either by experience or by religious psychology. As a result of our enquiry we must surely agree that we can set no limit to the ways through which God's presence may be felt, seeing that the Divine Spirit is striving to express itself in every circumstance and in every happening.

The error of the sacramentalist in the past has often rather been that he has confined the Divine presence and the Divine working to certain fixed channels and unchanging visible signs. We who hold that these good men have narrowed down the freedom of the inner life need to meet them, not by denying the Divine presence where they see it, but by trying to see and realise that presence ourselves more fully throughout all our lives.[1]

In all these things the real test is the life of man. "By their fruits ye shall know them."

However, there is one great way of approach confirmed by every kind of religious experience, supported by science and by the new work in psychology. This is the way of prayer. It is commonly affirmed that the scientist with his world of law and order and his evolutionary schemes, has banished prayer: to affirm this is to misread the later chapters of our story. In man we find the spiritual nature becoming the deciding factor; and in the spiritual world prayer is the opener-up of communications. Prayer is the vehicle or the means whereby the influence from God can enter into our lives. Our freedom and our control of things is so important a part of the purpose of God that even in spiritual matters God does not force His power upon us. We can go our own way and ignore the guiding hand of God—we are all conscious of having done so—but we can go God's way helped by the conscious guidance of His Spirit. Prayer is the great means, whereby God can perform in and through us His will for the world. We can regard prayer as the move, from man's side, which enables God, from His side, to do what He has been longing to do but could not, until man opened the way. The laws of the spirit are not broken by man's use of prayer as the way of inter-

[1] *A Wayfarer's Faith*, T. E. Harvey.

communication, any more than a bird breaks the laws of the physical world when flying overhead. The act of prayer can be likened to the act of flight. The two actions are alike in this, they exhibit the creative power in life to use the laws of the Universe. Just as a bird makes use of the air and the gravitational pull in its method of flight, so man makes use of the power of spiritual communion to draw strength from the source of all strength. This is not the place to go into a general discussion upon prayer—there are many difficulties which perforce are slurred over here—all that I wish to maintain is that the idea of prayer is natural and not unnatural, that its efficacy and value are not destroyed by our thesis of the winning of freedom. Indeed prayer is one of the chief ways in which God can come into the life of man.

> In such a kingdom—an organic fellowship of inter-related persons—prayer is as normal an activity as gravitation is in a world of matter. Personal spirits experience spiritual gravitation, soul reaches after soul, hearts draw toward each other. We are no longer in the net of blind fate, in the realm of impersonal force—we are in a love-system where the aspiration of one member heightens the entire group, and the need of one—even the least—draws upon the resources of the whole—even the Infinite. We are in actual Divine-human fellowship.[1]

We come now to the final problem of our enquiry. Does the revelation of Jesus of the personal redeeming love of God fit into the scheme? Can we say that God's dealings with man, which Jesus

[1] *The Double Search*, Rufus M. Jones, p. 92.

shows are personal dealings, are the natural outcome of the scheme, and that God has been always striving to deal personally with the Universe?

First of all, it seems to me that all we have said about the Providence of God and the naturalness of God's dealings in Nature, and His greater scope in man, are confirmed by the revelation of Jesus. At least there is nothing in the Gospel story which suggests any magical or non-natural way between God and man. The wide view of sacramentalism, which we have urged, was certainly the view of Jesus, for His life and teaching were based upon the natural and continual intercourse between the Divine and the human spirit. He did not condemn the Temple ritual, nor ignore the religious rites and ceremonies in use in his day; but the whole trend of his life and teaching was away from formalism and ritual to an aspect of religion in which man and God held intercourse like a son with a father, and in which all life's circumstances could be a means of grace. That the fellowship meal should be especially remembered, as one of the times of revelation, is very natural; what is not natural is the failure to see that Jesus was concerned with the sanctification of the whole of life. To harden one act of fellowship into a magical and special rite is to endanger the real revelation.

What concerns us here is the discovery made by Jesus Christ of the power of the Spirit and the conditions under which that power could be received, not as the occasional afflatus which men had recognised from all time, but as the very warp and woof of the web of man's social life, the daily food by which men, women, and children could be satisfied and empowered for the greater enterprise—dowered with the inward resourceful wisdom that can make the liberty of each harmonise with the good of all.[1]

Jesus also revealed the naturalness of prayer and meditation. One of the most certain remembrances of those who laboured with Jesus was His constant need for prayer and communion with God. It must have seemed strange to them, almost like a weakness, until they realised its wonderful significance, and they too learned to use the same method.

This constant conscious and loving fellowship with God which Jesus enjoyed is the last and surest proof of the Goodness of God, and a sign of His purpose for man.

Dare we go further? Can anything more be said to bring the revelation of Jesus into the whole scheme of evolution? I am deeply conscious of my inability to deal with this important problem, but I do not feel content to close this enquiry without some attempt, however inadequate, to express my thoughts a little more fully. In doing so I recognise their incompleteness, but the attempt to express them may help me to obtain a deeper vision of the mystery of Christ.

After all the mystery of Christ, in relation to God and to this world of endeavour, will not be solved by any thoery or by consistent statement; it will be solved by each of us in the loving service of God. Yet it may well be that the honest facing

[1] *The Spirit*, p. 64.

of difficulties and the attempt to see the unity of God the Father and God the Creator, will encourage us to push forward further experiments in life and in thought, believing that redemption does not come finally by knowledge but by revelation and by reconciliation.[1] It is just here that the fact of Jesus the Christ can help us by shewing us the need for redemption.

But all of this, it may be urged, is only saying that Jesus was one of the great revealers of God. Man is seeking, God is revealing, always through every circumstance and happening. Surely Jesus is in line with all the other seers and prophets of the Lord? Most certainly this is so, but does this exhaust the mystery? All history is a revealing, the whole evolutionary story is a revealing, beauty is a revealing, for God's spirit is in all and around all, and the meaning can be read in some measure through each and all of these various codes. The central idea we have tried to get in closing this enquiry is that God can be known through His handiwork in the world of Nature, and that we cannot think of this world apart from Him. Just as the soul of man uses the body to express its meaning, so Nature, as we have tried to see it, expresses the meaning of God. All discovery therefore is man's side of the problem of God's revelation. The discoveries of Jesus are therefore the natural channels of God's revelation.

[1] I wish to express my special indebtedness to two books: *Reconciliation and Reality*, W. F. Halliday, M.A.; *Grace and Personality*, John Oman, M.A.

We shall all agree that, although God's meaning can be expressed in Nature, it is more adequately expressed through human personality and the more perfect the human personality the more perfect the revelation. That is to say, the more awake the revealer is to the personal influence of God, the more perfect will be the revelation. In recognising that Jesus was the noblest man who has ever lived, and that His life speaks more truly than the life of any other man before or since, some may be content to rest here and so explain the authority of His message.

But is this a sufficient explanation? Let us consider once more the purpose of the scheme, of which man and Jesus are parts. We have seen that the winning of freedom is the great result of the evolutionary method, and have tried to show that freedom is necessary to the larger purpose—the winning of personality and its fulfilment in the loving gracious fellowship between man and God. Incidentally our enquiry has helped us to see that the winning of freedom explains both physical and moral evil; that the environment in which it is possible to make a wrong response, is the necessary school for the personal and individual conquest of the evil and the victory of good. God as a person cannot force His love upon man, but must work along the paths of influence and conviction, drawing man up into this loving relationship. This relationship depends upon both man and God; there is the double search and there are mutual contributions. But it may be remarked, and rightly so, that the ideas hinted at in these latter sentences are due to the revelation of Jesus. We are using His revelation to explain His revelation. This is inevitable; we cannot, even if we wished, get into the place where the life-revelation of Jesus can have no meaning or no influence. But for His revelation of God's Personality, we should be working and groping in the semi-darkness. Jesus has illumined all our darkness. It may well be that we have to realise the meaning of the Christ in our own lives before we can begin to find the larger meaning of the life of Jesus in relation to the world. Professor John Oman puts it thus:

> What, in short, is the good of looking for Christ Who was meek and lowly in heart, in the Gospels, when we should be certain not to recognise Him in our next-door neighbour? Till we believe in Him there, we cannot possibly believe in Him anywhere else.[1]

So too, it may well be, that the answers to our questions will only be found as we know more of the inward reconciling power of the living Christ in our own lives.

Why could not this living spirit of Christ be revealed to us directly? Why is it associated with a historic person—Jesus of Nazareth? The answer is found in the purpose. For man to realise and to understand the Perfect Personality of God, there is needed, so it seems to me, someone to come between man and God, not in the sense of a connecting wire without which no communication can pass, but as an interpreter, one who can

[1] *Grace and Personality*, p. 135.

explain and interpret the meaning of the symbols and signs.

The gradual recognition of God as Power might conceivably be accomplished by man from his side learning more and more of the laws of Nature and their application to human life, but this is not the purpose of God. The recognition of God as Love, although it will come gradually as the result of a double desire and effort, needs interpreting by a person. The problem of bringing together God, the Perfect Person and men, the struggling imperfect and limited persons, seems to require for its solution a human life in as perfect fellowship with God as it is possible for humanity to be. The interpreter must be a man, otherwise the problem is still unsolved—and the interpreter must by his perfect humanity reveal God in such a way as to show what Divinity in humanity means.

This is what Jesus did and still does. His revelation is not confined by any measure of history, but continues and performs the interpretation or revealing of God to all who desire it.

God's purpose explains the incarnation. Man learns from the life, death and resurrection of Jesus that human life can be lived divinely. He not only sees such a life, but in Jesus he finds the way, the method for raising his own life to this finer purpose.

Jesus is the sign as well as the fulfilment of God's purpose. This means that we get to a loving relation to God through Jesus; that by reason of His revelation we know our kinship with God;

that no great gulf divides human nature from Divine nature; and that for us Jesus has lead us on to the pathway that leads to sonship and to the Kingdom of God. I have already been under such obligation to Professor John Oman for one or two extracts, which have explained my meaning far better than I could have done, that I hesitate to use him again; but the following paragraphs do help in putting clearly the views I have been endeavouring to express.

Revelation, being thus concerned with the reconciliation to God's gracious relation to us by which alone we can discover that it is gracious, must be a work of history. What is more, it must be the work of history, the work which give it meaning and treasures up its gains. The life of everyone who takes the right road and uses life to the right end and lays his heart open to the right influences, will help to interpret God's gracious relation to his fellows as well as to himself. But there will be special significance in the experience of those who meet life with special insight and sincerity and courage, especially if they be exercised in times of supreme crisis in human affairs. As in all other human progress, they will establish one line of advance, so conspicuously in the right direction as to make all others mere matter of antiquarian interest. Finally, if there were One whose absolutely right relation to God manifested adequately God's relation to us, even that line would become only a preparation for His task, and He would be an ultimate revelation, not in the sense of being a substitute for our own insight or of exhausting the whole meaning of experience, but as the inspiration of our insight and the pioneer of our experience. Yet Christ is the supreme revelation only as He is the supreme reconciliation. Its finality is not as the guarantee of a body of truth which makes of no account God's patient wisdom in overcoming unbelief manifested in all human history, but as the embodiment of a relation to the Father, the perfection of which we prove only as we use it to interpret His relation to us in all things and at all times.

* * * * *

The fellowship of the Spirit is no mere emotional sense of God's presence or even of His power, but a personal dealing with God in His world as the Spirit of Holiness, to be realised in all the tasks and trials of life, in inheriting the earth as the manifestation of God's moral purpose with us, in enabling us, in a true moral society, to live in all sympathy with men, in all pardon of offences, and in all calm loyalty to every cause of righteousness.

* * * * *

The revelation of God, so understood, means that it belongs eternally to His nature not to be content to direct the world according to His own wise love, heedless of our misunderstanding, or to offer us His fellowship, heedless of our alienation, but that He must seek to overcome, in the freedom of a true reconciliation, our misunderstanding and our alienation. That is the end of all His dealings with us in time, and the task to which He has called all prophetic souls, and which is consummated in the Lord of the prophets, who, being perfectly the Son of God, enables us to be sons of God, for whom, in the fellowship of the spirit, all things work together for good.[1]

Continuity and unity, these are the two last words of this enquiry.

There is one God and he has been the same throughout all time. His love has been around us all the way. God the Creator is God the Father of Jesus and so is the Father of us all. God the Creator is God the Immanent spirit, Immanent in Nature, in man, in Jesus, the inner light that

[1] *Grace and Personality*, pp. 146, 150–1.

enlightens all our darkness, that sustains all life, that inspires all effort. Continuity and unity. God's Providence is one, and His love has made the world and His love draws all mankind unto Himself.

BOOKS

Those mentioned in the footnotes.

A PRAYER

OUR Father, we thank Thee for the revelation of Thyself in Nature and in Man, the child of Nature. Every bush aflame with beauty, every bird athrob with song, every man aware of the higher, one and all reveal Thy presence.

Open our eyes, we pray Thee, that we may miss neither the many-splendoured thing, nor the glory of Thy love in human fellowship, but read Thy message through the varied codes in Nature, and in Man. We thank Thee for the way we have been led, for Thy wondrous confidence throughout the long, long, upward road of toil; may we feel our debt to those whose struggles and endeavours have brought us here, and strive to be more worthy of Thy trust.

Help us to see Thy love in the world, even in the pain and anguish. Above all, we pray Thee, help us to understand Thy greatest revelation in the loving gracious relationship seen in Christ, whose earthly life showed Thy will of goodness as the meaning and purpose of life.

Help us to see Thy world as Thou seest it, to see Man as Thou seest him, for only so can we have confidence and hope.

May Thy Kingdom come in us. AMEN.

INDEX

PAGE

Antiquity of man . . . . 19 ff

Beauty as a code . . . . 129 ff
,, fact of . . . . 123 ff
,, what it is 124–5
Bell, Sir Charles 13
Bridgewater Treatises 13
Browning, R. 111

Darwin, C. 13, 25, 54
Darwinism 27
Driesch, H. 38, 42, 52
Drummond, H. 36, 37, 55
Drummond, W. 129

Entelechy 29, 42
Environment and Organism . . . . 47 ff, 60 ff
,, conquest of . . . . . . 61, 92 ff, 142
,, perfect . . . . 64 ff, 137, 150 ff
Evil, physical . . . . 103 ff
,, moral . . . . 108 ff
Evolution. . . . . 46 ff, 91 ff, 141 ff
,, and evil . . . . 103 ff
,, and experimentation . . . . 92 ff
,, and gregariousness . . . . 57 ff, 72 ff, 74 ff
,, and Jesus 171
,, Natural Selection 51
,, nature and nurture . . . . 74, 144 ff
,, of spiritual nature . . . . 33 ff
,, purpose of . . . . 66 ff, 166
,, recapitulation idea in 24
,, struggle for existence 51, 53
,, variation 50
,, web of life 51, 93
Evolutionary scheme as training ground . . . . 100 ff

PAGE

Fairless, M. 131
Freedom for goodness 98
,, the purpose of evolution . . . . . . 96 ff
,, the winning of . . . . 69 ff

Gregariousness . . . . 57, 72 ff
,, and religion . . . . 79 ff

Halliday, W. F. 150, 153, 166
Hand of man 14
,, homologies 14, 15
,, vestiges 16
,, the freed 28
Herd instincts . . . . 57, 72 ff
Huxley, T. H. 54

Immanence of God. . . . . 153 ff

Jefferies, R. . . . . 128, 133 ff
Jesus. 165, 171
Johnstone, J. 40, 41

Keith, A. 16, 24
Kropotkin, P. 55, 56

Lankester, E. R. 112, 148

Man, a moral being . . . . 99 ff
,, æsthetic activity in . . . . 124 ff, 137
,, and control of Nature 147
,, and religious experience . . . . . . 84 ff, 164
,, antiquity of 19
,, approach to God 168, 170
,, gregariousness of . . . . 74 ff
,, place in Nature . . . . 11 ff, 17
,, ,, embryological evidences . . . . . 20 ff
,, ,, environmental stimuli . . . . . 62 ff
,, ,, evolutionary factors . . . . . . 71 ff
,, spiritual nature of . . . . 32 ff
,, the purpose of evolution 71
,, uniqueness of 25, 30, 142
McDougall, W. 84, 85

PAGE
McDowall, S. A. 125, 138
Micklem, N. 31
Mitchell, P. Chalmers 84, 117

Nature and nurture . . . . 74, 144 ff

Oman, J. 162, 163, 176, 178
Organism and environment . . . . 47 ff

Pain . . . . 103 ff
Personality 65, 69, 167
Prayer 170
Providence of God . . . . 152, 158 ff

Religion and gregariousness . . . . 79 ff
Religious experience . . . . 84 ff, 164
Revelation of Jesus . . . . 171 ff
Royce, J. . . . . 119 ff
Ruskin, J. 127, 128, 132

Sin . . . . 107 ff

Tennant, F. R. 111
Thomson, J. Arthur . . . 2, 13, 18, 19, 43, 48, 49, 149, 150
Thoreau 89
Transcendence of God . . . . 153 ff
Trotter, W. . . . . 57, 71 ff, 76, 116
Troubetzkoy, E. 29, 86

Vestiges 16, 17
Vitalism . . . . 39 ff

Wallace, A. R. 12
War and evolution . . . . 112 ff
Wordsworth, W. 130